Chromatography/Fourier Transform Infrared Spectroscopy and Its Applications

PRACTICAL SPECTROSCOPY

A SERIES

Edited by Edward G. Brame, Jr.

The CECON Group
Wilmington, Delaware

Chromatography/Fourier Transform Infrared Spectroscopy and Its Applications

Robert White

Department of Chemistry
The University of Oklahoma
Norman, Oklahoma

CRC Press
Taylor & Francis Group
Boca Raton   London   New York

CRC Press is an imprint of the
Taylor & Francis Group, an **informa** business

First published 1990 by Marcel Dekker, INC

Published 2019 by CRC Press
Taylor & Francis Group
6000 Broken Sound Parkway NW, Suite 300
Boca Raton, FL 33487-2742

© 1990 by Taylor & Francis Group, LLC
CRC Press is an imprint of Taylor & Francis Group, an Informa business

First issued in paperback 2019

No claim to original U.S. Government works

ISBN 13:978-0-367-45088-5 (pbk)
ISBN 13:978-0-8247-8191-0 (hbk)

Visit the Taylor & Francis Web site at
http://www.taylorandfrancis.com

and the CRC Press Web site at
http://www.crcpress.com

Library of Congress Cataloging-in-Publication

White, Robert
    Chromatography/Fourier transform infrared spectroscopy and its
    applications / Robert White.
        p. cm. -- (Practical spectroscopy ; v. 10)
        ISBN 0-8247-8191-0
        1. Infrared spectroscopy. 2. Fourier transform spectroscopy.
    3. Chromatographic analysis. I. Title. II. Series.
    QD96.I5W47 1990
    543'.08583--dc20   89-23738
                                                        CIP

Preface

The combination of chromatographic separation and Fourier transform infrared spectroscopy (FT-IR) has proven to be a powerful analytical tool and has significantly increased capabilities for complex mixture analysis. This book is intended to serve as an up-to-date reference source for those familiar with chromatography/FT-IR methods and as an introduction to techniques and applications for those interested in future uses for chromatography/FT-IR.

The first chapter deals with general characteristics common to all chromatography/FT-IR interfaces. Necessary compromises in the marriage of separation methods with FT-IR detection are outlined. Subsequent chapters (2–4) describe the instrumental aspects of interfaces for the three classes of chromatographic separation: GC, HPLC, and TLC. Advantages and disadvantages of each interface method are described in detail. Current procedures for identification of separated mixture components based on infrared spectral characteristics are described in Chapter 5. In addition, uses of complementary information for "unequivocal" structure elucidation are discussed. Chapter

6 contains examples of chromatography/FT-IR applied to a variety of complex mixture analysis problems. Future trends and applications are predicted based on the focus of current research.

Robert White

Contents

Chromatography/Fourier Transform Infrared Spectroscopy and Its Applications

1
Chromatography/Fourier Transform Infrared Spectroscopy Interfaces

I. INTRODUCTION

The term "chromatography" was coined by Tswett in 1906 and is derived from the Greek words for color (khromatos) and write (graphos) [1]. In Tswett's experiment demonstrating chromatography, plant pigments were separated into color zones by passing plant extract dissolved in petroleum ether through a column containing calcium carbonate. Plant pigments were separated by differing adhesion to calcium carbonate particles. This was the first application of *adsorption chromatography*. *Partition chromatography* differs from adsorption chromatography in that separation is achieved by distributing solute between two immiscible phases instead of adhesion to a solid surface. Martin and Synge demonstrated partition chromatography in 1941 [2]. They separated mono-amino acids in protein hydrolysates by partition between chloroform and water phases. Water adsorbed on silica gel was used as the stationary phase. Methyl orange indicator was added to the aqueous phase in order to observe the movement of separated acids through the column. Martin and Synge suggested that vapor mobile phases

could be employed as well as liquids. Their work formed the
foundation for development of modern liquid and gas chromatography.

Chromatographic apparatus and methodologies have been
developed to facilitate analysis of mixtures containing hundreds
of components. A summary of currently available chromatographic
methods is given in Table 1-1. Component isolation can be
achieved for solutes contained in gas, liquid, or supercritical
phases. Methods employed for detecting separated components
must discriminate between chromatographic phases and isolated
mixture constituents. Many different discrimination methods
have been devised for in-situ detection of chromatographically
separated substances (Table 1-2). These detectors monitor
physical or chemical properties that change when mixture components are present. Bulk property chromatography detectors
provide little qualitative information but can be used for quantitative analysis when mixture component identities are known.
Substance identification requires implementation of detectors
that provide structural information for separated species. Infrared spectrometry is one such detection method. Infrared

Table 1-1   Classification of Chromatographic Methods

Classification	Mobile/stationary phase	Separation mechanism
GC	gas/liquid	partition
	gas/solid	adsorption
	gas/bonded phase	partition/adsorption
LC	liquid or supercritical	
	fluid/liquid	partition
	liquid/solid	adsorption/ion exchange/ sieving
	liquid or supercritical	
	fluid/bonded phase	partition/adsorption
TLC	liquid/solid	adsorption
	liquid/bonded phase	partition/adsorption

Table 1-2 Some Common Chromatography Detectors

Detector	Chromatographic method		
	GC	LC	TLC
Thermal conductivity	[3]	[4]	[5]
Flame ionization	[6]	[7]	[8]
Electron capture	[9]	[10]	—
Conductance	[11]	[12]	—
UV	[13]	[14]	[15]
Fluorescence	[16]	[17]	[18]
Infrared	[19]	[20]	[21]
Mass spectrometry	[22]	[23]	[24]
NMR	—	[25]	—

vibrational spectra provide information regarding structure-
dependent molecular motions. Vibrational spectroscopy can be
used to identify molecular functionalities and is particularly
useful for isomer discrimination.

II. FOURIER TRANSFORM INFRARED SPECTROSCOPY

Fourier transform infrared spectroscopy (FT-IR) is the preferred method for infrared detection of chromatographically
separated species. The popularity of FT-IR is primarily due
to the multiplex and rapid scanning features of interferometry.
The interferometer was developed by Albert A. Michelson during the 1880s and was described in a paper published in 1891
[26]. Interferometers used in FT-IR instruments manufactured
today are similar in design to the one built by Michelson.
Figure 1-1(a) is a diagram of a two-beam Michelson interferometer. Interferometry produces a complex waveform that is a sum
of contributions from all wavelengths emitted by the source.
Wavelength discrimination derives from the property that wavelength contributions are modulated at different frequencies. In
contrast, a diffraction grating disperses source radiation and

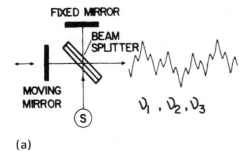

FIXED MIRROR
BEAM SPLITTER
MOVING MIRROR
S
ν_1, ν_2, ν_3

(a)

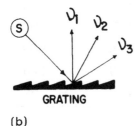

S
ν_1 ν_2
ν_3
GRATING

(b)

Figure 1-1   (a) A two-beam Michelson interferometer.   (b) Wavelength dispersion by a diffraction grating.

each wavelength occupies a different position in space [Figure 1-1(b)]. Fundamental distinctions between FT-IR and dispersive IR are evident in mathematical descriptions. For example, wavelength resolution for infrared monochromators incorporating diffraction gratings is described by

$$n\lambda = d\{\sin(i) + \sin(r)\} \tag{1.1}$$

where n is the diffraction order, d the spacing between grating grooves, i the angle of incidence of infrared radiation to the grating, and r the angle of diffraction selected for detection. The spacing between grating grooves is fixed and cannot be altered without changing the grating. In Eq. (1.1), λ is described as a function of spatial coordinates determined by the angles i and r.

The mathematical expression for interferometric wavelength discrimination is the Fourier transformation

$$I(\nu) = \int_{-\infty}^{\infty} I(t) \; e^{(-i2\pi\nu t)} \; dt \tag{1.2}$$

where $i = \sqrt{-1}$, $I(\nu)$ is the intensity of frequency ν, and $I(t)$ is the amplitude of the interferogram at time t. Equation (1.2) contains no spatial parameters but instead reflects temporal wavelength dependence associated with interferometry. In practice, Eq. (1.2) is not explicitly solved. Instead, a finite approximation known as the fast Fourier transformation (FFT) is employed [27]. A detailed description of Fourier transform infrared spectroscopy is beyond the scope of this book. However, the primary distinctions between interferometry-based FT-IR and wavelength dispersion will be described briefly. Interested readers are encouraged to consult books by Bell [28], and Griffiths and de Haseth [29] for an in-depth treatment of the subject.

A. Fellgett's Multiplex Advantage

Fellgett's multiplex sensitivity advantage derives from the fact that all wavelengths (resolution elements) are detected simultaneously by FT-IR [30]. To explain the multiplex advantage, it is helpful to consider an FT-IR spectrometer as being equivalent to an array of monochromators, each adjusted to detect one resolution element of the complete infrared spectrum. As illustrated in Figure 1-2(a), n monochromators are required to simultaneously measure n resolution elements of an infrared spectrum. All monochromator components in the array are identical; only the orientation of each grating (i.e., the wavelength detected) differs. In comparison, a dispersive spectrometer measures resolution elements sequentially [Figure 1-2(b)]. In both methods, the time spent measuring each resolution element determines the signal-to-noise ratio (SNR) for that wavelength in the spectrum. If we assume that each resolution element is measured for a time t_{RE}, the time required to obtain a complete spectrum by using dispersive spectroscopy would be

$$t_{tot} = N_{RE} t_{RE} \tag{1.3}$$

where N_{RE} is the number of resolution elements measured. In comparison, the time required to obtain a comparable spectrum by using a monochromator array (FT-IR) would be t_{RE}. Thus,

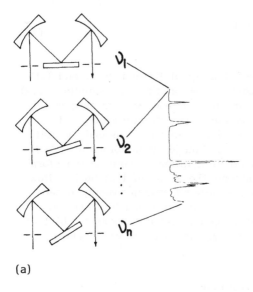

(a)

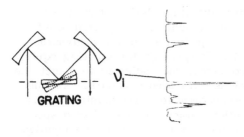

(b)

Figure 1-2   (a) FT-IR represented as an array of monochromators.   (b) A dispersive monochromator.

for the same SNR, FT-IR provides a savings in measurement time equal to N_{RE}.

If FT-IR measurements are made for the same length of time required for dispersive spectrum recording, the SNR for the FT-IR (monochromator array) measurement should exceed that of the dispersive spectrometer. The SNR for a measurement made by an infrared detector is proportional to the square root of the measurement time. For FT-IR measurements made during the time period required for dispersive spectrum recording, FT-IR spectral SNR_{FT} would be proportional to $(N_{RE}t_{RE})^{1/2}$. In contrast, the spectral SNR for a dispersive infrared spectrometer measurement in the same time period (SNR_D) would be proportional to $t_{RE}^{1/2}$. The improvement in SNR for the FT-IR measurement compared to the dispersive measurement would be given by

$$\frac{SNR_{FT}}{SNR_D} = \frac{(N_{RE}t_{RE})^{1/2}}{t_{RE}^{1/2}} = N_{RE}^{1/2} \tag{1.4}$$

For spectral resolution elements sampled at 1 cm^{-1} intervals over the mid-infrared region (4000 cm^{-1}–400 cm^{-1}), Fellgett's advantage predicts a factor of 60 improvement in SNR for FT-IR vs wavelength dispersion for identical measurement periods.

B. Jacquinot's Throughput Advantage

Wavelength dispersion requires the use of a monochromator that incorporates entrance and exit slits. These slits reduce the radiant energy transmitted through the spectrometer. In contrast, an FT-IR has no slits. Therefore, FT-IR power transmission is greater than for a dispersive spectrometer operating at comparable resolution. Increased radiant power transmission by FT-IR compared to dispersive infrared spectrometers is known as Jacquinot's throughput advantage [31,32]. The magnitude of Jacquinot's advantage is wavelength-dependent and can vary from about 20 to 200 [33].

Hirschfeld has described the advantages of using a low throughput FT-IR for high-sensitivity applications requiring small-diameter infrared beams. This "counter-Jacquinot" or "Hirschfeld" sensitivity advantage derives from the use of small area detectors that generate less noise than larger detectors [34]. Using a 50 μm diameter detector, Hirschfeld obtained

subnanogram detection limits for infrared analysis of gas
chromatographic eluents [34].

C. Connes' Advantage (Wavelength Accuracy)

A diffraction grating ruled with periodic groove spacing errors
will produce ghosts at the monochromator focal plane that degrade spectral resolution. Similar ghosts can be observed in
FT-IR spectra caused by errors in the measurement of interferometer moving mirror displacement [35]. These artifacts are
effectively eliminated by employing monochromatic (laser) interferometry to measure mirror displacements with high precision.
Typical wavelength accuracy for an FT-IR equipped with a
632.8 nm He/Ne laser reference is approximately 0.01 cm^{-1}.

D. Stray Light Effects

Monochromators are sensitive to stray light interferences and
must be enclosed to eliminate external polychromatic radiation.
An FT-IR detector responds to intensity modulations produced
by an interferometer. Although stray light does contribute to
detector saturation in FT-IR, it is not modulated and therefore
does not contribute to spectral intensities. However, detector
saturation can produce spectral artifacts and should be avoided.

E. Detector

Not all of the distinctions between FT-IR and wavelength dispersion are in favor of interferometry. Sensitive thermocouple
detectors commonly employed in dispersive instruments are not
appropriate for rapid scanning interferometry. The noise
equivalent power (NEP) for thermocouple detectors is approximately 1×10^{-10} W $Hz^{-1/2}$ at 15 Hz chopping frequencies commonly employed for dispersive infrared spectroscopy [33]. The
NEP for pyroelectric DTGS detectors used in FT-IR is about 20
times greater than a thermocouple detector under these conditions. Thus, some of the improvement derived from Fellgett's
and Jacquinot's advantages is negated by detector insensitivity.
However, overall performance is generally in favor of FT-IR for
chromatographic analysis applications.

III. CHROMATOGRAPHY/FT-IR INTERFACE OPTIMIZATION

An ideal interface between chromatography and FT-IR would couple both techniques without sacrificing the performance of either. Specific interface designs depend on many factors, including the physical state of matter to be analyzed. Common interfaces for infrared analysis of vapors, liquids, and powders are shown in Figure 1-3. Vapor phase GC/FT-IR measurements require long-path-length transmission cells [Figure 1-3(a)]. Short-path-length flow cells are used in liquid chromatography (HPLC/FT-IR) to minimize absorbance interferences from infrared opaque mobile phases [Figure 1-3(b)]. Diffuse reflectance is the most efficient method for FT-IR analysis of solutes deposited on highly scattering stationary phases (TLC/FT-IR) [Figure 1-3(c)]. Chapters 2—4 contain more details about specific chromatography/FT-IR interfaces.

A. Chromatographic Separation Criteria

All chromatographic separation methods comprise a mobile phase and a stationary phase. For separation to occur, mixture constituents must have differing affinities for these phases. The partition coefficient (K_p) is a measure of the relative affinity of an analyte (A) for the stationary phase relative to the mobile phase in partition chromatography.

$$K_p = \frac{[A]_S}{[A]_M} \qquad (1.5)$$

Partition coefficients for all mixture components must be different if complete chromatographic separation is to be achieved. Partition coefficients for chromatographic solutes depend on the nature of solute interactions with mobile and stationary phases, as well as separation conditions.

Chromatographic methods can be classified into two general categories: column and planar (Figure 1-4). In column chromatography, mixture components are passed through a tube containing stationary phase. Due to partition coefficient

(a)

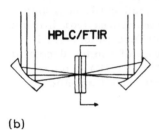

(b)

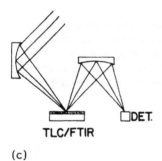

(c)

Figure 1-3   Typical optical arrangements employed for detection of (a) vapor phase (b) liquid phase, and (c) solid phase chromatographic phases.

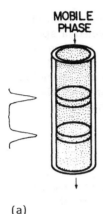

(a)

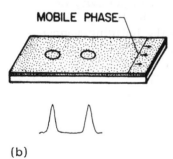

(b)

Figure 1-4 (a) Column chromatography. (b) Planar chromatography.

differences, some components are retained within the tube for longer periods than others. As a result, the time required for individual components to exit the column varies for each component separated [Figure 1-4(a)]. Gas chromatography (GC) and liquid chromatography (LC) are column chromatography separation methods.

In planar chromatography, the same mechanisms active in column chromatography are responsible for mixture component isolation. In contrast to column chromatography, stationary phase is contained on an open, flat surface instead of inside a tube, and component removal from the stationary phase is not

achieved during separation. Instead, mobile phase is allowed
to pass over the stationary phase until it has traveled a specified distance. During this time, the distance that mixture components deposited on the stationary phase move is inversely
proportional to their adsorption affinity. Component separation
is reflected by the positions of individual components on the
flat surface relative to the distance traveled by the mobile
phase [Figure 1-4(b)].

1. Column Chromatography

Van Deemter et al. derived an equation relating experimental
parameters to column chromatographic performance in 1956 [36].
This equation adequately described column chromatographic
phenomena as it was known at the time. A modified version of
the van Deemter equation is more applicable to column chromatography as it exists today [37]. Column dispersivity or
"height equivalent to a theoretical plate" (h) [2] can be related to chromatographic parameters in the following manner:

$$h = \frac{2\gamma D_m}{\mu} + \frac{qkd_f^2 \mu}{(1+k)2D_s} + \frac{\text{funct}(d_p^2, d_c^2, \text{end}, \mu)\, \mu}{D_m} \qquad (1.6)$$

where:

μ = linear mobile phase velocity

D_m = molecular diffusion coefficient for solute in mobile
phase

D_s = molecular diffusion coefficient for solute in stationary
liquid phase

d_p = support-particle diameter

d_c = column diameter

d_f = stationary liquid phase film thickness

k = column capacity ratio
(K_p × stationary phase volume/mobile phase volume)

q = liquid phase distribution constant

γ = obstruction factor

end = column end-effects

From Eq. (1.6), it is apparent that the efficiency of column chromatographic separations is dependent on many factors. Column diameter (d_c), mobile phase flow rate (μ), and column end-effects (end) are important considerations for developing FT-IR interfaces. Mobile phase flow rate determines the minimum FT-IR scan repetition rate required to preserve chromatographic resolution. Greater separation efficiency (smaller h) is obtained by using small-diameter columns. This is especially true for gas chromatography (e.g., capillary gas chromatography). Column end-effects are included in the last term of Eq. (1.6). This implies that the chromatography /FT-IR connection may have an effect on chromatographic resolution. Therefore, the size of the chromatographic column must be considered when devising methods for attaching the column to the FT-IR interface.

2. Planar Chromatography

Mathematical models describing dispersivity (h) for thin-layer chromatography are less accurate than for column chromatography. This is due to a variety of reasons, but primarily because of the fact that stationary phase is in contact with both liquid and vapor mobile phase during separation. Mobile phase vapor can alter stationary phase adsorption characteristics, resulting in unpredictable dispersivity changes during separation. In spite of this, a mathematical model has been developed that relates experimental parameters for planar chromatography to separation efficiency [38]

$$h = \frac{2\gamma D_m (z_1 + z_A)}{K_m R_F} + \frac{3 A d_p^{5/3} (z_1^{2/3} - z_A^{2/3})}{2(2D_m)^{1/3} (z_1 - z_A)} \tag{1.7}$$

where

z_A = distance that solute moves relative to solvent level

z_1 = distance that mobile phase moves relative to solvent level

K_m = mobile phase velocity constant

A = plate stationary phase packing quality factor

R_F = retardation factor

Retardation factor (R_F) is an experimentally derived parameter

$$R_F = \frac{z_s}{z_m} \tag{1.8}$$

where z_s is the distance that solute travels during separation and z_m the distance that mobile phase travels during separation relative to the initial solute position.

It is important to note that Eq. (1.7) has no dependence on mobile phase flow rate. Instead, dispersivity is determined by the distance that the mobile phase travels. Like column chromatography, separation efficiency for planar chromatography is dependent on the particle size of the stationary phase (d_p). Stationary phase particle size and solute elution distance are the most important considerations for the design of an FT-IR interface for planar chromatography. Stationary phase particle size determines the degree to which infrared radiation is scattered from the chromatographic surface. Solute elution distance determines the surface area that must be analyzed by FT-IR.

B. FT-IR Performance

Instrumental and experimental factors that affect FT-IR measurements have been discussed in detail elsewhere [28,29]. Only those factors that are important for chromatography/FT-IR measurements will be discussed here.

1. Interferometer

Applications of FT-IR coupled with column chromatography often require rapid interferometer scan rates in order to avoid an apparent degradation of chromatographic resolution. Interferometer scan velocities are limited by the digitization rate of the analog-to-digital converter (ADC) employed. This limitation currently restricts interferometer scan rates to 3.5 cm/sec (110 kHz ADC rate). However, a 16-bit, 200 kHz ADC is now commercially available that should eventually increase interferometer scan rates to 6 cm/sec.

Capillary gas chromatography (GC) is the column separation technique requiring the most rapid data acquisition. Capillary GC solute elution times between 5 and 10 sec can be observed for minor mixture components. Intrinsic band widths for vapor phase infrared absorptions measured by GC/FT-IR are on the order of $10-20$ cm^{-1}. Thus, an infrared spectral resolution of 8 cm^{-1} or 4 cm^{-1} is adequate for routine GC/FT-IR measurements. At 4 cm^{-1} spectral resolution, an interferometer scan rate of 1 cm/sec produces approximately four interferograms per second. Thus, a 1 cm/sec scan rate is sufficient to characterize the sharpest capillary gas chromatographic elutions containing enough material to be detected by GC/FT-IR. For example, capillary GC elutions with peak widths of 5 sec would be digitized by 20 infrared spectra at 4 cm^{-1} resolution by employing a 1 cm/sec interferometer scan velocity. As stated previously, current maximum interferometer scan rates are more than three times faster (3.5 cm/sec).

2. FT-IR Stability

Although minimum interferometer scan rates are readily attained, there is a need for improved interferometer stability. Unstable interferometers can cause excessive noise and baseline drift in spectra acquired during lengthy separations. De Haseth has characterized short- and long-term instabilities for a rapid scan interferometer [39]. He found that sources of interferometer instability were difficult to identify but may include such things as timing errors, vibrations, source instability, power fluctuations, laser instability, interferometer velocity fluctuations, detector fluctuations, and mechanical stress relaxation.

A rigorous test of interferometer stability consists of monitoring the reproducibility of interferograms measured over a period of time [40]. In theory, subtraction of successive interferograms should reveal detector and electronic amplification noise only. Typically, a residual interferogram is also observed. The maximum of the residual does not usually coincide with the original interferogram maximum. Instead, it is often associated with interferogram regions in which the slope of the waveform is steep [near the zero path difference point (ZPD)]. The magnitude of subtraction residuals is indicative of the repeatability of interferometer scans. Figure 1-5 illustrates typical interferogram subtraction results for FT-IR. Figure 1-5(a) is a representative single-scan interferogram. Figure 1-5(b) is an interferogram subtraction result showing complete removal of interferogram information, leaving only detector and amplifier noise. Figure 1-5(c) is an average interferogram residual from 25 interferogram subtractions. A small interferogram subtraction residual is clearly visible near ZPD in Figure 1-5(c). Note that the ordinate scales of interferogram residuals shown

REPRESENTATIVE SINGLE SCAN INTERFEROGRAM

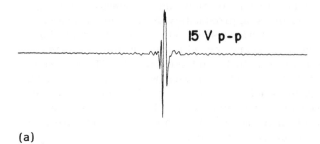

(a)

INTERFEROGRAM SUBTRACTIONS

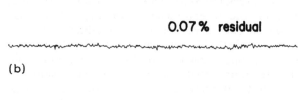

(b)

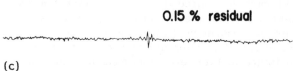

(c)

Figure 1-5   (a) Representative single-scan FT-IR interferogram.   (b) Interferogram subtraction result showing complete removal of wavelength information.   (c) Interferogram subtraction result exhibiting an interferogram residual.   (Reprinted with permission from Ref. 40. Copyright © 1985 Society for Applied Spectroscopy, Frederick, Maryland.)

in Figures 1-5(b) and 1-5(c) are expanded by a factor of 100 relative to a representative interferogram.

3. Detector

The most common detectors employed for FT-IR mid-infrared measurements are the deuterated triglycine sulfate (DTGS) pyroelectric detector and the mercury-cadmium-telluride (MCT) liquid-nitrogen-cooled semiconductor alloy detector. Nearly all chromatography /FT-IR measurements requiring rapid data acquisition have employed MCT detectors. MCT detectors are more sensitive than DTGS detectors and have faster response times. In fact, liquid-nitrogen-cooled MCT detector response is rapid enough to permit operation at data-acquisition rates above 1 MHz [41]. DTGS detectors are less expensive than MCT detectors and can be employed for applications in which high interferometer scan rates are not required.

The interferometer scan rate required for optimum FT-IR performance is dictated by the dependence of detector sensitivity on modulation frequency and scan duty cycle efficiency. The SNR for FT-IR measurements is inversely proportional to the square root of the interferometer scan rate [Eq. (1.9)]. This is reflected in Figure 1-6, which is a plot of spectral

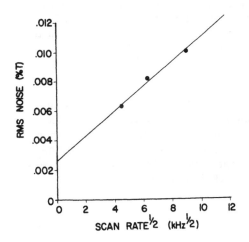

Figure 1-6   FT-IR spectral noise plotted as a function of interferometer scan velocity.

noise vs the square root of scan velocity for a fixed number of signal-averaged scans.

FT-IR SNR is directly proportional to the square root of the number of signal-averaged scans [Eq. (1.10), Figure 1-7].

$$SNR \propto v^{-1/2} \qquad\qquad (1.9)$$

$$SNR \propto n_s^{1/2} \qquad\qquad (1.10)$$

Equations (1.9) and (1.10) indicate that increases in SNR obtained by signal averaging more scans at a higher scan rate should be offset by a decrease in SNR resulting from operating at the higher scan rate. However, as shown in Figure 1-8, MCT detector sensitivity (detectivity D*) increases with modulating frequency (interferometer scan rate). Therefore, a SNR improvement would be expected. Unfortunately, interferometer scan duty cycle efficiency decreases with increasing scan rate in a manner that negates detector sensitivity increases (Figure 1-9). As a result, little dependence of SNR on interferometer scan rate is observed.

A problem long associated with MCT detectors is their apparent ease of saturation and nonlinear response. Both of these problems can result in spectral artifacts [42]. Hirschfeld recently proposed that observed nonlinearities are caused by nonlinear detector preamplifiers and not the detectors [34]. Efforts to improve preamplifier design and eliminate these artifacts are currently under way [43].

C. Interface Evaluation

Performance of chromatography/FT-IR interfaces depends on the tradeoffs made in chromatographic resolution and FT-IR detection. In general, three parameters can be used to characterize chromatography/spectroscopy interfaces. Resolution degradation (R_d) represents the loss of chromatographic resolving power incurred by the FT-IR interface.

$$R_d = \frac{R_{CHROM}}{R_{FT-IR}} \qquad\qquad (1.11)$$

In Eq. (1.11), R_{FT-IR} is the chromatographic resolution as determined from a chromatogram derived from FT-IR measurements

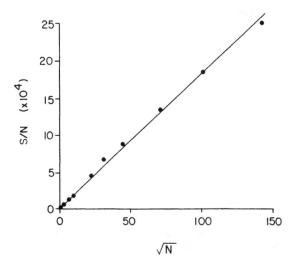

Figure 1-7 FT-IR SNR plotted as a function of the number of signal-averaged scans.

and R_{CHROM} is the chromatographic resolution as it would have been measured in the absence of the FT-IR interface. Ideally, R_d would be unity. In practice, R_d is always greater than unity. An FT-IR interface should be designed in a manner that minimizes R_d.

The concept of an enrichment factor [44] is applicable to systems in which one or both chromatographic phases can be

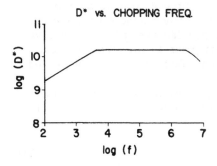

Figure 1-8 Variation of D* as a function of modulation frequency for a liquid-nitrogen-cooled MCT detector.

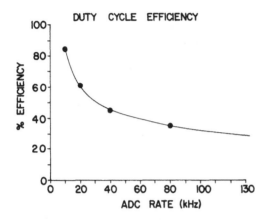

Figure 1-9   Interferometer scan duty cycle efficiency as a
function of interferometer scan velocity.

present during solute detection and these phases give rise to
spectral interferences or solute dilution.   The enrichment
factor (E) for an analyte (A) can be defined as

$$E = \frac{[A]_{FT-IR}}{[A]_{CHROM}}$$ (1.12)

where $[A]_{FT-IR}$ is the concentration of analyte in the interfer-
ing phase(s) during infrared detection and $[A]_{CHROM}$ is the
concentration of the analyte in the chromatographic system
prior to FT-IR analysis.   Obviously, when spectral interferences
can be attributed to the chromatographic system, it is best to
remove the solute from the system prior to infrared analysis.
For these systems, large enrichment factors result in improved
interface performance.

The yield (Y) of a chromatography/spectroscopy interface
is the percentage of solute ultimately detected

$$Y = \frac{Q(A)_{FT-IR}}{Q(A)_{CHROM}} \times 100\%$$ (1.13)

where $Q(A)_{FT-IR}$ is the quantity of analyte detected by the FT-IR and $Q(A)_{CHROM}$ is the quantity of analyte separated. The yield for an ideal interface would be 100%. Many chromatography/FT-IR interfaces approach this limit. However, some interfaces do not recover all of the analyte originally contained in the sample, resulting in yields below 100%.

IV. CHROMATOGRAM GENERATION FROM INFRARED DATA

A chromatogram is a plot of detector response as a function of separation time or distance and is used to represent mixture-component resolution. Chromatogram peaks indicate the presence of separated components. The property monitored for separated component detection by FT-IR is infrared absorbance. Unfortunately, it is not obvious how to extract infrared absorbance information from time domain interferogram data generated by FT-IR [see Figure 1-5(a)]. A solution to this problem would be to employ Fourier transformation [Equation (1.2)] to compute infrared spectra for each acquired interferogram and measure infrared absorbance directly from absorbance spectra. This necessitates repeated Fourier transformations during chromatographic separation and may not be feasible for interfaces requiring rapid data acquisition. In these instances, chromatographic resolution may be sacrificed if a Fourier transformation cannot be performed quickly. Alternatively, chromatograms can be generated from interferogram data by monitoring changes in the shape of digitized time domain waveforms. In general, methods for monitoring interferogram changes are not equivalent to measurement of infrared absorbance but are useful for detecting separated components and can be used for quantitative analysis.

Several chromatography/FT-IR chromatogram construction methods were evaluated by Hanna et al. [45]. They investigated methods based on direct interferogram comparisons, as well as those requiring Fourier transformation prior to data evaluation. Table 1-3 contains a summary of their comparison for some chromatogram generation schemes. The Gram—Schmidt vector orthogonalization method was twice as sensitive as any of the other techniques and required fewer floating point operations than methods employing fast Fourier transformation (FFT).

Table 1-3   Comparison of Some GC/FT-IR Chromatogram
Generation Methods

Method	No. floating point Operations	SNR
ΣI_n^2	200	1.5
Euclidean distance	320	9.3
Gram—Schmidt orthogonalization (5 basis vectors)	1230	8.7
Gram—Schmidt orthogonalization (30 basis vectors)	6200	23.0
Absorbance integration (64-point FFT)	2370	5.5
Absorbance integration (256-point FFT)	13,000	9.5

Source: Reproduced from Ref. 45 by permission of Preston
Publications, a division of Preston Industries, Inc., Niles, Illinois.

They concluded that the Gram—Schmidt vector orthogonalization
method was the most sensitive and computationally most efficient
of those tested.   Subsequent developments in computer technology made rapid computation of low-resolution infrared absorbance spectra feasible and permitted real-time integrated absorbance measurements for gas and liquid chromatographic
separations.   White et al. compared the performance of Gram—
Schmidt chromatogram reconstruction with integration of 16 cm^{-1}
(1024-point FFT) absorbance spectra [46].   They also found
that the Gram—Schmidt method was more sensitive for recon-
structing chromatograms than Fourier transformation followed
by absorbance integration.   However, they reported that absorbance integration could be used to generate functional group-
specific chromatograms with greater SNR than either the total
absorbance or Gram—Schmidt orthogonalization methods.   Unfortunately, functional group-specific chromatograms must be
computed at the expense of increased calculation time.   Initially,
this was a significant drawback because available computers

were relatively slow. This problem has been circumvented with the development of array processors and multiprocessing computer technologies.

A. Interferogram-Based Methods

The Gram—Schmidt orthogonalization process is a method used by mathematicians to generate coordinate systems in multidimensional space [47]. The technique was first applied to GC/FT-IR chromatogram reconstruction by de Haseth and Isenhour in 1977 [48]. For Gram—Schmidt chromatogram generation, a portion of each interferogram acquired during separation is selected to represent a multidimensional vector. The number of interferogram data points used to form vectors determines the dimensionality of the vector space and hence the number of orthogonal axes needed to completely define the space. With the Gram—Schmidt process, a subspace is derived from interferograms representing mobile phase (and possibly, stationary phase) absorptions. This vector subspace is highly representative of the background vectors that were used to construct the subspace. To a close approximation, each background vector can be resolved into a linear combination of subspace axes. After the subspace is specified, chromatogram intensity values are defined as the length of the residual vector that remains after vector components within the background subspace are removed from chromatogram vectors. For background interferogram vectors, this residual is small. However, for interferograms containing absorbance information, residuals can be large.

The Gram—Schmidt chromatogram reconstruction method is illustrated for three-dimensional space in Figures 1-10 and 1-11. The basis set (background vector subspace) is formed from background interferograms prior to the start of chromatographic separation. One background (reference) interferogram vector ($\vec{R}_1$) is arbitrarily selected to designate the first subspace axis [Figure 1-10(a)]. The $\vec{R}_1$ vector is comprised of d sequential interferogram data points (r_i), where d designates the dimensionality of the vector space.

$$\vec{R} = (r_1, r_2, \ldots, r_d) \tag{1.14}$$

The first basis vector ($\vec{B}_1$) is formed by normalizing the reference vector ($\vec{R}_1$) to unit length [Figure 1-10(b)].

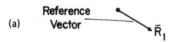

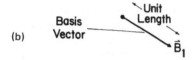

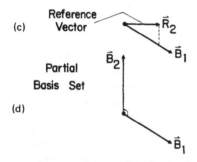

Figure 1-10   The Gram–Schmidt method for partial basis set formulation.   (a) Arbitrarily selected reference vector.   (b) Normalization of (a) to produce a basis vector.   (c) Construction of a vector orthogonal to the first basis vector from a second reference vector.   (d) Normalization of the orthogonal vector to produce a second basis vector and partial basis set.

$$\vec{B}_1 = \frac{\vec{R}_1}{\sqrt{\vec{R}_1^T \, \vec{R}_1}} \tag{1.15}$$

where

$$\vec{R}_1^T \, \vec{R}_1 = \sum_{i=1}^{d} (r_d)^2 \tag{1.16}$$

A partial basis set is constructed as follows: A second reference vector $(\vec{R}_2)$ is used to construct a new vector $(\vec{O})$ that is orthogonal to $\vec{B}_1$ [Figure 1-10(c)].

$$\vec{O} = \vec{R}_2 - (\vec{B}_1^T \vec{R}_2)\vec{B}_1 \qquad (1.17)$$

The orthogonal vector $(\vec{O})$ is normalized to create a second basis vector $(\vec{B}_2)$ [Figure 1-10(d)].

$$\vec{B}_2 = \frac{\vec{O}}{\sqrt{\vec{O}^T \vec{O}}} \qquad (1.18)$$

At this point [Figure 1-10(d)], a partial basis set (two of three possible axes) has been formed for the three-dimensional space and chromatogram intensities can be calculated. For construction of larger partial basis sets (i.e., when $d > 3$), additional basis vectors can be obtained by

$$\vec{O}_k = \vec{R}_k - \sum_{i=1}^{k-1} (\vec{B}_i^T \vec{R}_k)\vec{B}_i; \qquad k < d \qquad (1.19)$$

$$\vec{B}_k = \frac{\vec{O}_k}{\sqrt{\vec{O}_k^T \vec{O}_k}} \qquad (1.20)$$

Figure 1-11 illustrates chromatogram intensity calculations employing the two-dimensional Gram–Schmidt-generated subspace [Figure 1-10(d)]. Successive interferogram-derived vectors obtained during separation $(\vec{C}_J)$ are resolved into basis set components and orthogonal residuals are computed.

$$\vec{O}_R = \vec{C}_J - \sum_{i=1}^{n} (\vec{B}_i^T \vec{C}_J)\vec{B}_i; \qquad n < d \qquad (1.21)$$

In Eq. (1.21), n is the number of basis vectors used to define the background subspace. The value of n must necessarily be

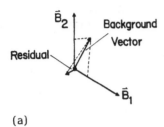

(a)

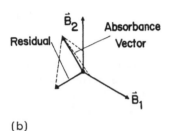

(b)

Figure 1-11   Chromatogram generation using a Gram—Schmidt partial basis set.   (a) Orthogonal residual computed from a background vector.   (b) Orthogonal residual computed from a vector containing solute absorbance information.   (c) Gram—Schmidt GC/FT-IR chromatogram.

less than d, the dimensionality of the space.   If n were equal to d, all chromatogram vectors could be represented as a linear combination of basis vectors and all residuals would be zero.

Orthogonal residuals ($\vec{O}_R$) for vectors derived from interferograms not containing solute information are small because these interferograms are almost completely comprised of linear combinations of basis vectors [Figure 1-11(a)].   Solute infrared absorptions result in interferogram shape distortions that alter both the orientation and length of extracted vectors.   As a result, orthogonal residuals computed for vectors derived from interferograms containing solute absorbance information are larger than those obtained from background interferograms [Figure 1-11(b)].   A plot of orthogonal residual length as a function of chromatographic separation time or distance produces a chromatogram [Figure 1-11(c)].

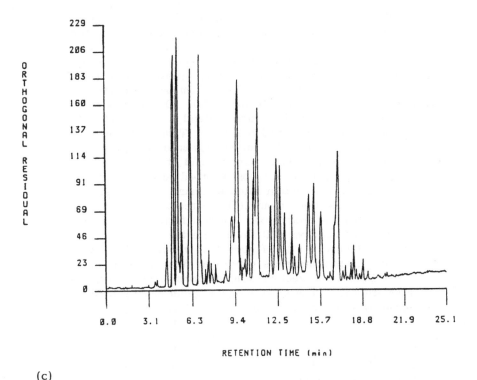

(c)

Figure 1-11 (Continued)

From Table 1-3, it is apparent that the number of basis vectors used to define the background vector subspace has a significant effect on the SNR of generated chromatograms. The portion of interferograms used to create vectors also affects the SNR of generated chromatograms. In their original paper, de Haseth and Isenhour empirically determined the interferogram region that produced the best SNR for two different chromatograms [48]. Their optimum vector was extracted beginning 60 data points from the interferogram zero path difference point (ZPD) and extended for 100 data points (100 dimensions). White et al. used simplex optimization [49] to find the optimum interferogram region for forming vectors [46]. They concluded, contrary to what de Haseth and Isenhour had found, that the

highest SNR was obtained when data points near ZPD were
used to form vectors. Subsequent reports indicate that the
choice of optimum interferogram vector displacement is dependent on the nature of separated components and the stability of the interferometer employed for measurements [50—52].

Time domain oscillations resulting from broad spectral absorptions are located primarily near ZPD. Oscillations corresponding to sharp spectral features extend farther past ZPD
than those for broad spectral features. This phenomenon is
illustrated in Figure 1-12 for interferogram data derived from
vapor phase measurements of pentane and ethyl acetate. Pentane infrared spectra contain mostly broad features, whereas
ethyl acetate spectra contain intense, sharp features (Figure
1-13). Vector selections near ZPD are required to properly
weight contributions from broad spectral features. Vector
selections displaced from ZPD favor detection of sharp spectral
absorptions (Figure 1-12). Figure 1-14 shows GC/FT-IR
Gram—Schmidt chromatograms generated from the same data but
using 100-dimensional vectors sampled from different regions of
the interferogram. The chromatogram generated by using
interferogram information in the region of ZPD [Figure 1-14(b)]
compares favorably with total absorbance integration computed
by Fourier transforming each interferogram, generating absorbance spectra, and computing integrated absorbance [Figure
1-14(a)]. Figure 1-14(d) is a chromatogram generated by
selecting vectors displaced 120 data points from ZPD. The
SNR of chromatogram peaks for pentane and heptane is significantly diminished in Figure 1-14(d) compared to Figure 1-14(b),
whereas the SNR for ethyl acetate is relatively unchanged.
Consequently, ethyl acetate is the largest component in the
chromatogram. From these observations, a vector sampling
containing ZPD information would be expected to produce optimum chromatogram SNR. However, Brissey et al. have shown
that noise contributed by interferometer instability must also
be considered in selecting the optimum interferogram region
for vector sampling [51]. They contend that interferogram
noise consists of three different components. Detector noise is
distributed evenly along the entire interferogram. Noise out-
side of the spectral bandwidth is also independent of position
in the interferogram. A third noise source is the instability of
the interferometer. When successive interferograms are subtracted, the largest residual always occurs near ZPD [Figure
1-15(c)]. This indicates that the effect of interferometer instability is greatest near ZPD. For unstable interferometers,

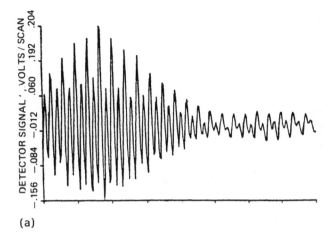

(a)

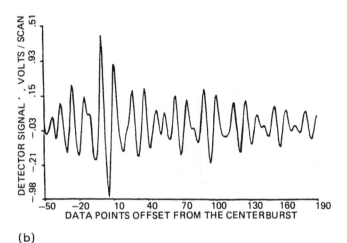

(b)

Figure 1-12 Absorbance interferograms derived by subtracting interferograms containing sample information for (a) pentane and (b) ethyl acetate from a background interferogram. (Reprinted with permission from Ref. 51. Copyright © 1984 American Chemical Society, Washington, D.C.)

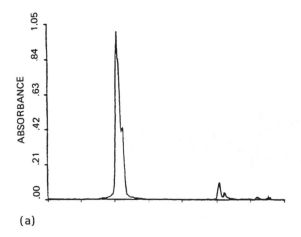

(a)

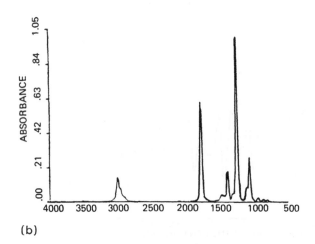

(b)

Figure 1-13   Vapor phase infrared absorbance spectra for (a)
pentane and (b) ethyl acetate.   (Reprinted with permission
from Ref. 51.   Copyright © 1984 American Chemical Society,
Washington, D.C.)

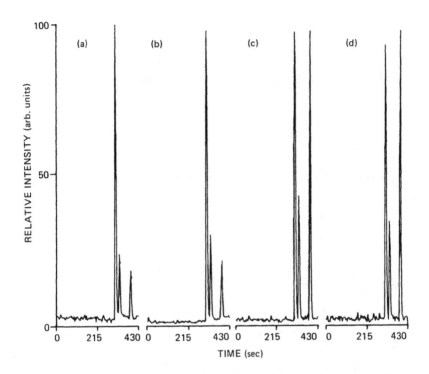

Figure 1-14   Comparison of GC/FT-IR chromatograms for a separation of pentane, heptane, and ethyl acetate.   (a) Total integrated absorbance.   (b) Gram—Schmidt reconstruction based on a 100-data point vector sampled symmetrically about ZPD. (c) Gram—Schmidt reconstruction with 100-dimensional vectors displaced 60 data points from ZPD.   (d) Gram—Schmidt reconstruction with 100-dimensional vectors displaced 120 data points from ZPD.   (Reprinted with permission from Ref. 50.   Copyright © 1983 American Chemical Society, Washington, D.C.)

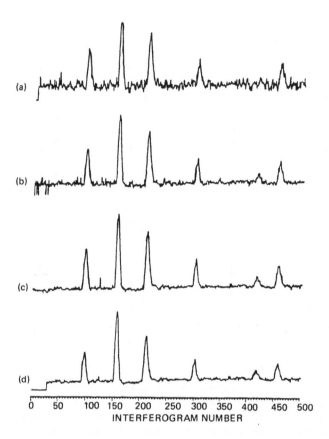

Figure 1-15   Chromatogram reconstructions for a mixture of
bis(2-chloroethyl) ether, acetophenone, methyl salicylate,
2,3,5-trimethylphenol, acenaphthene, and 2,4,6-trimethylphenol.
(a) 5 Basis vector Gram−Schmidt reconstruction.   (b) 5 Basis
vector modified Gram−Schmidt reconstruction.   (c) 5 Eigen-
vector factor analysis reconstruction.   (d) 30 Basis vector
Gram−Schmidt reconstruction.   (Reprinted with permission
from Ref. 54.   Copyright © 1982 American Chemical Society,
Washington, D.C.)

the SNR of chromatograms generated by using information near ZPD is low due to large noise contributions from the interferometer in this region. Recent reports propose an acceptable vector sampling for all interferometers beginning at 10 data points past ZPD [51,52]. FT-IR interferometer stability has minimal effect on the SNR of reconstructed chromatograms when this sampling is employed.

The potential for developing computationally efficient chromatogram reconstruction procedures based on algorithms such as Gram—Schmidt orthogonalization has inspired attempts to refine vector techniques. Rice has shown that a more appropriate basis set can be formed with a modification of the Gram—Schmidt orthogonalization technique [53]. In this procedure, potential new basis vectors are rejected if orthogonal residuals ($\vec{O}_K$) are too small. This method has been applied to GC/FT-IR chromatogram generation to enhance chromatogram SNR [54]. Owens et al. have used principal component factor analysis [55] to identify significant features of interferogram-derived vectors and create a 5 vector basis set that performed as well as a 30 basis vector Gram—Schmidt chromatogram reconstruction [54] [Figures 1-15(c) and 1-15(d)]. The 5 basis vectors were derived from 30 background interferograms by using principal component factor analysis. Real-time chromatogram generation was possible with the 5 basis vector Gram—Schmidt method, whereas the time required for orthogonalization to 30 basis vectors was prohibitive for capillary GC/FT-IR applications.

B. Functional Group-Specific Chromatograms

Coffey et al. introduced functional group-specific chromatograms (chemigrams^{TM}) in 1978 [56]. Chemigrams are produced by computing absorbance spectra during chromatographic separation and plotting integrated absorbance for prespecified wavelength windows as a function of separation time. Alternatively, pattern-recognition methods have recently been applied to produce functional group-specific chromatograms without computing absorbance spectra [57]. Functional group chromatograms are analogous to GC/MS mass chromatograms and represent the quantity of a specific functional group that elutes as a function of time (column chromatography) or distance (planar chromatography). As previously stated, construction of functional group-specific chromatograms is not as computationally efficient as the Gram—Schmidt orthogonalization technique but provides specific information regarding structures of separated mixture components.

Functional group-specific integrated absorbance chromatograms are generated by summing infrared absorbances in pre-
defined spectral windows during chromatographic separation.
Absorbance integration necessitates the computation of absorbance spectra for each acquired interferogram and requires
hardware and software to perform rapid Fourier transformations.
A comparison of chromatographic peak SNR measured for four
separated components (toluene, p-xylene, heptane, o-dichlorobenzene) by using different spectral windows is contained in
Table 1-4. Optimum SNRs for the first three components were
obtained for windows in the C—H stretching region. This was
expected for toluene, p-xylene, and heptane. Little absorbance
was measured in the aliphatic C—H stretching region for the
fourth peak, which was consistent with its identity (o-dichlorobenzene). The best window for detection of peak 4 was the
$1070-970$ cm^{-1} region. Vapor phase absorbance spectra of
o-dichlorobenzene contain absorbance maxima at 750 cm^{-1}. This
absorbance band was not included in any of the windows
selected. Obviously, effective use of absorbance windows requires some knowledge of mixture composition. It is noteworthy that the total absorbance window ($4000-800$ cm^{-1} in
Table 1-4) always yielded lower chromatographic peak SNR than
optimum functional group-specific reconstructions.

The absorbance integration procedure amplifies baseline
noise within selected spectral windows when integrated absorbance values are computed. Bowater et al. [58] have improved the SNR of functional group-specific chromatograms by
employing the maximum absorbance algorithm [59]. Chromatogram intensity obtained by using the maximum absorbance algorithm is defined as the largest absorbance within the specified spectral window. The effect of noise is *not* compounded by
using the maximum absorbance algorithm as it is for absorbance
integration. Figure 1-16 is a comparison of GC/FT-IR chromatograms for a mixture of (in order of elution) 1,1,2-trichloro-
ethane, o-xylene, o-dichlorobenzene, and nitrobenzene obtained
by using integrated and maximum absorbance algorithms.
Figure 1-16(a) is a chromatogram generated by integrating absorbance over the region $771-677$ cm^{-1}. Figure 1-16(b) is a
chromatogram generated by using the maximum absorbance algorithm for the same wavelength window. For all separated
components, chromatographic peak SNR obtained with the maximum absorbance algorithm was at least twice that obtained by
integrated absorbance.

Table 1-4   Peak SNR for Different IR Integration Windows

Peak	SNR for IR Windows (cm^{-1})				
	4000–800	3130–3010	3000–2930	1070–970	850–780
Toluene	4.6	3.4	55.0	2.6	2.8
p-Xylene	3.0	12.0	8.7	4.3	2.2
Heptane	3.1	8.6	11.6	3.7	6.8
o-Dichlorobenzene	2.3	4.7	2.2	6.6	3.4
Average	3.3	7.2	19.4	4.3	3.8

Source: Reprinted with permission from Ref. 46. Copyright ©
1981 American Chemical Society, Washington, D.C.

C.   Quantitative Analysis

Both integrated absorbance and Gram—Schmidt vector methods
have been used for the quantitative analysis of mixture components.   Hembree et al. have applied GC/FT-IR to the quantitative analysis of polycyclic aromatic hydrocarbons (PAHs)
[60].   They obtained a detection limit of 710 ng for pyrene.
A linear Beer's law plot of absorbance vs quantity of pyrene
was obtained to a limit of 8 µg.   Hembree et al.'s results also
compared favorably with GC/MS analysis.   For example, phenol
and o-cresol were detected in shale oil at levels of 340 and
370 ppm, respectively, by GC/FT-IR.   Corresponding concentrations obtained by GC/MS were 334 and 322 ppm, respectively.
   Sparks et al. [61] were able to quantify amounts of pentyl
propionate over a range from 461 ng to 138 µg from Gram—
Schmidt-generated GC/FT-IR chromatograms.   A linear relationship between chromatographic integrated peak area and concentration was obtained for on-column quantities of pentyl propionate between 461 ng and 10 µg.   However, Malissa et al.
found that plots of integrated Gram—Schmidt peak area vs concentration were nonlinear at low concentrations [62].   They attributed this nonlinearity to spectral baseline drift during
chromatographic separation that, at low concentration, was of

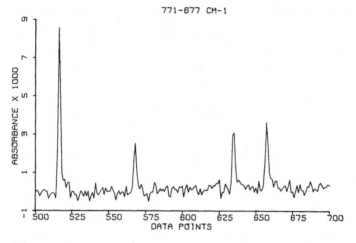

(a)

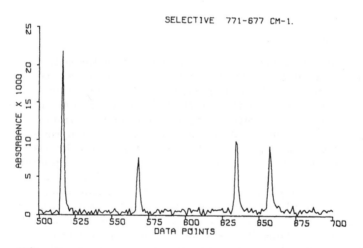

(b)

Figure 1-16 Comparison of functional group-specific chromatograms constructed using (a) integrated absorbance and (b) maximum absorbance algorithms. (Reprinted with permission from Ref. 58. Copyright © 1986 American Chemical Society, Washington, D.C.)

the same magnitude as sample absorption. Lam et al. demonstrated that improved Gram—Schmidt quantitative analysis could be obtained by cross-correlating chromatographic peaks with a Gaussian function prior to area integration [63]. Cross-correlation was used to preserve chromatographic peak area while at the same time providing chromatogram smoothing to reduce noise. Cross-correlation smoothing was found to be better than simply increasing the number of basis vectors employed in the Gram—Schmidt orthogonalization process.

The GIFTS software package developed by Hanna et al. greatly simplified GC/FT-IR data manipulations and has served as a template for commercially available chromatography/FT-IR software packages [64]. The GIFTS package provided for chromatogram reconstruction, spectrum selection, and automated library searching [65]. Chromatography/FT-IR software for the same functions is now commercially available from a number of FT-IR instrument manufacturers.

V. SUMMARY

Chromatography/FT-IR combinations have proven to be very useful for the structural analysis of complex mixture components. To fully appreciate chromatography/FT-IR interface tradeoffs, one must understand how chromatographic efficiency and infrared spectroscopy are affected by the method of connecting the two instruments. Three interface parameters defined in this chapter—resolution degradation (R_d), enrichment factor (E), and yield (Y)—are used in subsequent chapters to compare and contrast different methods of coupling chromatographic separations with infrared detection.

REFERENCES

1. M. Tswett, *Ber. Debut. Bot. Ges.*, *24*: 234 (1906).

2. A. J. P. Martin and R. L. M. Synge, *Biochem. J.*, *35*: 1358 (1941).

3. A. E. Lawson and J. M. Miller, *J. Gas Chromatogr.*, *4*: 273 (1966).

4. K. Okzeki, T. Kambara, and K. Saitoh, *J. Chromatogr.*,
    38: 393 (1968).

5. E. Haahti and I. Jaakonmaki, *Ann. Med. Biol. Exp. Fenn.*,
    47: 175 (1969).

6. I. G. McWilliam and R. A. Dewar, *Gas Chromatography*
    (D. H. Desty, ed.), Academic Press, New York, pp. 142–
    152 (1958).

7. A. T. James, J. R. Ravenhill, and R. P. W. Scott, *Chem.*
    Ind., *18*: 746 (1964).

8. T. Cotgreave and A. Lynes, *J. Chromatogr.*, *30*: 117
    (1967).

9. J. E. Lovelock and S. R. Lipsky, *J. Am. Chem. Soc.*, *82*:
    431 (1960).

10. G. Nota and R. Palombari, *J. Chromatogr.*, *62*: 153
    (1971).

11. O. Piringer and M. Pascalau, *J. Chromatogr.*, *8*: 410
    (1962).

12. A. J. P. Martin and S. S. Randall, *Biochem. J.*, *49*: 293
    (1951).

13. M. Novotny, F. I. Schwende, M. J. Hartigan, and
    J. E. Purcell, *Anal. Chem.*, *52*: 736 (1980).

14. C. G. Horvath and S. R. Lipsky, *Nature*, *211*: 748
    (1966).

15. R. Markham and J. D. Smith, *Nature*, *163*: 250 (1949).

16. R. P. Cooney and J. D. Winefordner, *Anal. Chem.*, *49*:
    1057 (1977).

17. M. Toporak and L. J. Phillip, *J. Chromatogr.*, *20*: 299
    (1965).

18. D. Jerchel and W. Jacobs, *Angew. Chem.*, *65*: 342 (1953).

19. A. M. Bartz and H. D. Ruhl, *Anal. Chem.*, *36*: 1892
    (1964).

20. K. L. Kizer, A. W. Mantz, and L. C. Bonar, *Am. Lab.*,
    7 (5): 85 (1975).

21. C. J. Percival and P. R. Griffiths, *Anal. Chem.*, *47*:
    154 (1974).

22. J. C. Holms and F. A. Morrell, *Appl. Spectrosc.*, *11*: 86 (1957).

23. M. A. Baldwin and F. W. McLafferty, *Org. Mass Spectrom.*, 7: 1111 (1973).

24. R. Kaiser, *Chem. Br.*, *5*: 54 (1969).

25. E. Bayer, K. Albert, M. Nieder, E. Grom, and T. Keller, *J. Chromatogr.*, *186*: 497 (1979).

26. A. A. Michelson, *Phil. Mag. Ser.*, *31*: 256 (1891).

27. J. W. Cooley and J. W. Tukey, *Math. Comput.*, *19*: 297 (1965).

28. R. J. Bell, *Introductory Fourier Transform Spectroscopy*, Academic Press, New York (1972).

29. P. R. Griffiths and J. A. de Haseth, *Fourier Transform Infrared Spectrometry*, Wiley-Interscience, New York (1986).

30. P. Fellgett, *J. Phys. Radium*, *19*: 187 (1958).

31. P. Jacquinot and C. J. Dufour, *J. Rech. C.N.R.S.*, *6*: 91 (1948).

32. P. Jacquinot, *Rep. Prog. Phys.*, *23*: 267 (1960).

33. P. R. Griffiths, H. J. Sloane, and R. W. Hannah, *Appl. Spectrosc.*, *31*: 485 (1977).

34. T. Hirschfeld, "1985 International Conference on Fourier and Computerized Infrared Spectroscopy" (J. G. Grasselli and D. G. Cameron, eds.) *Proc. SPIE*, *553*: 64 (1985).

35. P. Connes, *Lasers and Light*, Freeman, San Francisco, Calif. (1969).

36. J. J. van Deemter, F. J. Zuiderweg, and A. Klinkenberg, *Chem. Eng. Sci.*, *5*: 271 (1956).

37. S. J. Hawkes, *J. Chem. Ed.*, *60*: 393 (1983).

38. A. M. Siouffi, F. Bressolle, and G. Guiochon, *J. Chromatogr.*, *209*: 129 (1981).

39. J. A. de Haseth, *Appl. Spectrosc.*, *36*: 544 (1982).

40. R. L. White, *Appl. Spectrosc.*, *39*: 320 (1985).

41. *The Infrared Handbook* (W. L. Wolfe and G. J. Zissis,
     eds.), The Infrared Information and Analysis Center,
     Environmental Research Institute of Michigan, Ann Arbor,
     Michigan, p. 11-85 (1978).

42. D. B. Chase, *Appl. Spectrosc.*, *38*: 491 (1984).

43. R. A. Schindler, *NASA Tech. Brief*, *10* (2): Item #52
     (1986).

44. M. A. Grayson and C. J. Wolf, *Anal. Chem.*, *39*: 1438
     (1967).

45. D. A. Hanna, G. Hangac, B. A. Hohne, G. W. Small,
     R. C. Wieboldt, and T. L. Isenhour, *J. Chromatogr.
     Sci.*, *17*: 423 (1979).

46. R. L. White, G. N. Giss, G. M. Brissey, and C. L.
     Wilkins, *Anal. Chem.*, *53*: 1778 (1981).

47. J. W. Dettman, *Introduction to Linear Algebra and Differ-
     ential Equations*, McGraw-Hill, New York, p. 124 (1974).

48. J. A. de Haseth and T. L. Isenhour, *Anal. Chem.*, *49*:
     1977 (1977).

49. M. W. Routh, P. A. Swartz, and M. B. Denton, *Anal.
     Chem.*, *49* 1422 (1977).

50. R. L. White, G. N. Giss, G. M. Brissey, and C. L. Wilkins,
     Anal. Chem., *55*: 998 (1983).

51. G. M. Brissey, D. A. Henry, G. N. Giss, P. W. Yang,
     P. R. Griffiths, and C. L. Wilkins, *Anal. Chem.*, *56*:
     2002 (1984).

52. D. T. Sparks, P. M. Owens, S. S. Williams, C. P. Wang,
     and T. L. Isenhour, *Appl. Spectrosc.*, *39*: 288 (1985).

53. J. R. Rice, *Math Comp.*, *20*: 325 (1966).

54. P. M. Owens, R. B. Lam, and T. L. Isenhour, *Anal.
     Chem.*, *54* 2344 (1982).

55. E. R. Malinowski and D. G. Howery, *Factor Analysis in
     Chemistry*, Wiley, New York (1980).

56. P. J. Coffey, D. R. Mattson, and J. C. Wright, *Am.
     Lab.*, *10* (5): 126 (1978).

57. B. A. Hohne, G. Hangac, G. W. Small, and T. L. Isen-
     hour, *J. Chromatogr. Sci.*, *19*: 283 (1981).

58. I. C. Bowater, R. S. Brown, J. R. Cooper, and
    C. L. Wilkins, *Anal. Chem.*, *58*: 2195 (1986).

59. A. C. J. H. Drouen, H. A. H. Billiet, and L. De Galan,
    Anal. Chem., *57*: 962 (1985).

60. D. M. Hembree, A. A. Garrison, R. A. Crocombe,
    R. A. Yokley, E. L. Wehry, and G. Mamantov, *Anal.
    Chem.*, *53*: 1783 (1981).

61. D. T. Sparks, R. B. Lam, and T. L. Isenhour, *Anal.
    Chem.*, *54*: 1922 (1982).

62. H. Malissa, K. Martin, and K. Winsauer, "1985 Interna-
    tional Conference on Fourier and Computerized Infrared
    Spectroscopy" (J. G. Grasselli and D. G. Cameron, eds.)
    Proc. SPIE, *553*: 340 (1985).

63. R. B. Lam, D. T. Sparks, and T. L. Isenhour, *Anal.
    Chem.*, *54* 1927 (1982).

64. A. Hanna, J. C. Marshall, and T. L. Isenhour, *J.
    Chromatogr. Sci.*, *17*: 434 (1979).

65. D. F. Gurka and L. D. Betowski, *Anal. Chem.*, *54*: 1819
    (1982).

2
Gas-Liquid Chromatography/
Fourier Transform Infrared Spectroscopy

I. INTRODUCTION

Volatile mixture component analysis can be facilitated by employing gas-liquid chromatography (GC) to isolate mixture constituents. Detection of separated components is achieved by monitoring changes in physical or chemical properties of chromatographic effluent during component elution. Bulk property detectors (e.g., thermal conductivity, flame ionization, etc.) can be used for quantitative analysis when mixture components are known but provide little structural information for unknown mixture components. In contrast, infrared spectroscopy is a multidimensional detector that can provide a wealth of molecular structure information in addition to component detection. Vapor phase infrared spectra contain functional group-specific features and are particularly useful for isomer discrimination.

II. BRIEF HISTORY

The potential for obtaining structure-specific information for
substances separated by gas chromatography was the impetus
behind the first efforts to combine gas chromatography with
infrared spectroscopy. Initially, separated components were
isolated from GC effluent as condensed fractions and analyzed
by using a microcell or attenuated total reflectance spectroscopy. Later, vapor phase eluents were trapped inside long-
path-length gas cells and subsequently analyzed by using dispersive IR spectrometers [1–5]. Gas chromatograph/infrared
spectrometer (GC/IR) interfaces permitting real-time eluent
analysis were built in 1964 by groups at Dow Chemical Co. [6],
Wilks Scientific Corp., and Esso Research and Engineering Co.
[7]. The Dow Chemical Co. interface consisted of a rapid
scan dispersive infrared spectrometer and a polished metal
light pipe. Rapid scanning was attained by using two infrared
grating monochromators. Infrared source radiation passed
through a 1/8 in. × 1/2 in. × 12 in. metal light pipe containing GC effluent and was alternately transmitted to the two
monochromators by a beam chopper (Figure 2-1). One of the
monochromators covered infrared wavelengths from 2.5 to 7 μm,
whereas the other spanned the 6.5 to 16 μm range. By using
dual monochromator scanning, complete GC/IR spectra could be
obtained in 16 sec. This was adequate for many packed column
gas chromatography applications.

Tsuda et al. described a novel GC/IR interface in which
vapor phase eluent was condensed and analyzed in a liquid
cell [8]. GC effluent was combined with CCl_4 vapor and
passed through a condenser. Separated mixture components
were condensed from the GC carrier gas stream along with CCl_4
solvent. Infrared analysis was performed by passing condensed
material through a 17 μl, 0.2 mm path length flow cell that was
positioned within the sample compartment of an infrared
spectrometer.

Low and Freeman recognized the potential advantages of incorporating interferometry in GC/IR interfaces and built the
first GC/FT-IR system in 1967 [9]. Unfortunately, the sensitivity of FT-IR at that time was poor compared to other GC
detectors. As a result, fraction collecting was still preferred
over on-line GC/FT-IR analysis in the late 1960s and early
1970s despite the fact that several linked GC/IR analysis systems had been reported. Advances in electronics and computer
technology during the 1970s had a dramatic impact on FT-IR

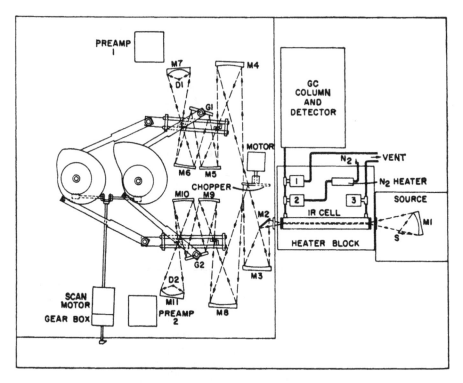

Figure 2-1   Optical layout of a dual-monochromator GC /IR
interface. (Reprinted with permission from Ref. 6. Copyright
© 1964 American Chemical Society, Washington, D.C.)

performance. Increased sensitivity, scan rate, data storage
capabilities, and availability of new methods for fabricating
light pipes resulted in more competitive GC /FT-IR detection
limits. These advances transformed GC /FT-IR from a research
curiosity to an affordable and effective tool for packed column
gas chromatographic analysis. Subsequent advances in light
pipe geometry optimization, interface optics, and infrared detectors led to the development of capillary GC /FT-IR. The sensitivity of GC /FT-IR was further increased by eluent trapping
techniques developed in the late 1970s and early 1980s. Matrix
isolation trapping of gas chromatographic eluents was first
demonstrated in 1979 by Bourne et al. [10]. In GC /MI /FT-IR,

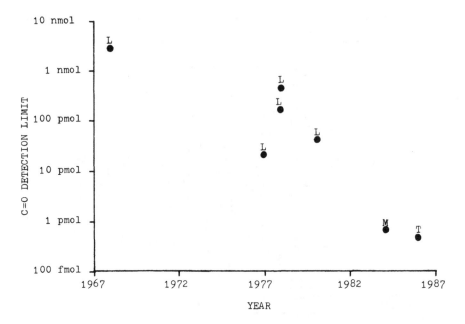

Figure 2-2 Reported GC/FT-IR detection limits since 1967. Points labeled L are from light pipe measurements, M are from matrix isolation, and T are from subambient trapping. Values were obtained from spectra contained in [12—18].

eluents are frozen in an inert argon matrix at 10°K. Matrix isolation offers a significant increase in GC/FT-IR sensitivity compared to light pipe interfaces. More recently, the concept of trapping GC eluents has been extended to include subambient freezing of mixture components [11]. In this interface, separated components are frozen on infrared transmitting materials and spectra are obtained by infrared microscopy.

GC/FT-IR detection limit improvements since 1967 are depicted by the graph in Figure 2-2. All detection limits shown were estimated from published spectra by measuring the SNR of carbonyl absorption bands and estimating the quantity of material that would have yielded a SNR of 2:1. Over the 20-year span shown, GC/FT-IR detection limits improved more than 1000-fold.

III. LIGHT PIPE INTERFACE

The first GC /FT-IR interfaces were composed of heated flow-through gas cells or *light pipes*. In a light pipe interface, infrared radiation passes through a highly polished tube that also transports gas chromatographic effluent (Figure 2-3). When mixture components elute, their presence is detected by the increased absorption of infrared radiation. If we utilize rapid scanning FT-IR, complete infrared spectra can be obtained for each vapor phase eluent passing through the light pipe.

A. Light Pipe Geometry

1. Optimum Volume

A light pipe contains no chromatographic stationary phase and does not contribute to mixture component separation. In fact, a light pipe and connecting tubing constitute a mixing volume in which separated components might recombine. This mixing results in chromatographic resolution degradation (R_d). Resolution loss is greatest for large "dead" volumes. Obviously, light pipe volumes should be reduced in order to minimize this effect. Unfortunately, if light pipe volumes are too small, little infrared radiation is transmitted and GC /FT-IR detection limits are unacceptably high. For optimum results, the best compromise between chromatographic resolution degradation and GC /FT-IR sensitivity must be found.

Conventional gas chromatographic detectors are designed to minimize dead volume in order to preserve chromatographic resolution. GC detectors commonly have dead volumes that are less than one-tenth of typical peak elution volumes. However, GC /FT-IR light pipe dead volumes must be significantly larger than conventional detectors in order to attain adequate sensitivity. The effect of light pipe dead volume on FT-IR-generated chromatograms can be illustrated by considering the fate of a Gaussian GC elution profile. A Gaussian elution can be represented by the following expression:

$$f(x) = (2\pi)^{-1/2} \sigma^{-1} e^{-x^2/2\sigma^2} \tag{2.1}$$

where σ is the standard deviation of the distribution (Figure 2-4). When $\sigma = 1$, the distribution is a standardized, centered,

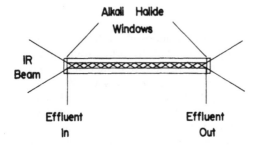

Figure 2-3   GC/FT-IR light pipe interface.

normal distribution.   This special case is noteworthy because
the integrated area of the Gaussian function is unity.   If we
use a Gaussian function as a model for component elution profiles, the fraction of total eluent (Q_{LP}/Q_{max}) and relative
analyte concentration ($[A]_{LP}/[A]_{max}$) can be computed as a
function of interface dead volume.   Figure 2-5 is a plot of
these parameters as a function of light pipe volume V_{LP}

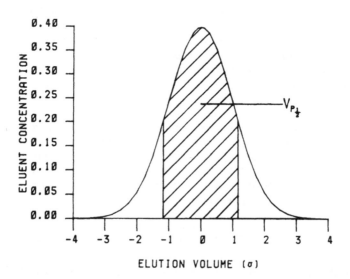

Figure 2-4   Gaussian GC elution profile.

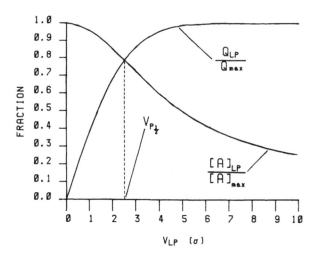

Figure 2-5 Plots of analyte fraction and relative concentration in a light pipe as a function of light pipe volume.

(in units of σ) for a Gaussian elution. The fraction of analyte detected increases with increasing light pipe volume, whereas the analyte concentration decreases. Infrared absorbance is proportional to analyte concentration and therefore also decreases with analyte dilution. The accepted compromise between fraction and concentration of analyte occurs at the intersection of the two curves. Light pipe volume at this point ($V_{P\frac{1}{2}}$) is equal to the volume of the GC elution contained within the full-width-at-half-maximum (FWHM) points of the distribution shown in Figure 2-4. The shaded area in Figure 2-4 constitutes 76.2% of the total elution volume.

The effect of dead volume on chromatographic resolution for a light pipe with a volume of $V_{P\frac{1}{2}}$ is illustrated in Figure 2-6. The two curves denote eluent concentration as a function of chromatographic elution volume. The curve marked "GC PEAK" represents the distribution of eluent as it exits the chromatographic column. The curve marked "GC/FT-IR PEAK" denotes a theoretical detected eluent profile that would be obtained by using a light pipe having a volume equal to $V_{P\frac{1}{2}}$. The detected GC/FT-IR distribution is broadened relative to the GC distribution. Chromatographic resolution degradation depicted in

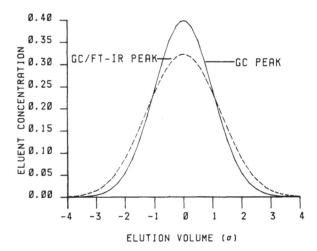

Figure 2-6 Chromatographic resolution degradation resulting from light pipe dead volume. Solid curve represents elution profile exiting chromatograph. Dashed curve represents detected elution profile as measured by GC/FT-IR.

Figure 2-6 is unavoidable for light pipe GC/FT-IR interfaces. The magnitude of this degradation can be estimated by considering the basic formula for calculating the resolution of two chromatographic elutions (denoted A and B)

$$R = \frac{2[(t_R)_B - (t_R)_A]}{W_A + W_B} \qquad (2.2)$$

where $(t_R)_A$ and $(t_R)_B$ are the retention times for the two elutions and W_A and W_B the peak widths. If we assume that peak widths are approximately equal, Eq. (2.2) becomes

$$R = \frac{[(t_R)_B - (t_R)_A]}{W} \qquad (2.3)$$

where $W = W_A = W_B$. For Gaussian profiles, $W = 4\sigma$. From Eq. (1.11), resolution degradation (R_d) can be written as

$$R_d = \frac{R_{CHROM}}{R_{FT-IR}} = \frac{\sigma_{FT-IR}}{\sigma_{CHROM}} \qquad (2.4)$$

When $V_{LP} = V_{P\frac{1}{2}}$, $\sigma_{FT-IR} = 1.2\sigma_{CHROM}$ and $R_d = 1.2$. The theoretical enrichment factor [Eq. (1.12)] for a light pipe with a volume $V_{LP} = V_{P\frac{1}{2}}$ can be estimated by the ratio of GC /FT-IR peak maximum to GC peak maximum (see Figure 2-6). The analyte concentration decreases due to dead volume mixing. Therefore, the enrichment factor in this case is less than unity ($E = 0.81$).

2. *Length/Diameter Considerations*

In the previous section, it was shown that optimum light pipe volume is directly related to the width of typical chromatographic elutions. For the most part, gas chromatographic elution widths are determined by the choice of chromatographic column (i.e., packed, SCOT, WCOT, etc.). For a given chromatographic system, an infinite number of length/diameter combinations exist that will provide a specified light pipe volume. For best GC /FT-IR performance, light pipe geometry must be optimized. Multiple reflections inside the light pipe increase beam path length when light pipe length is increased. Therefore, sample absorbance increases with increasing light pipe length if the volume is held constant [19,20] (Figure 2-7). However, transmittance is low for long light pipes because apertures are small. As a result, spectral noise increases with increasing light pipe length for fixed volume cells (Figure 2-8). A plot of spectral SNR for the 1240 cm^{-1} absorbance band of ethyl acetate as a function of light pipe length for a 0.5 ml volume light pipe is shown in Figure 2-9. The highest SNR was obtained for a 15 cm light pipe. For shorter light pipes, lower SNRs were obtained because path lengths were short, resulting in low absorbance. At lengths greater than 15 cm, noise caused by reflection and vignetting losses predominate, resulting in decreased 1240 cm^{-1} band SNR.

The transmittance of a light pipe has been related to its geometry by an empirical equation of the form [20]:

$$T_p = 0.9 \frac{d_p^2}{d_f^2} 0.32^{(3L/50 \, d_p)}; \qquad d_p/d_f = 1 \text{ when } d_p > d_f$$

$$(2.5)$$

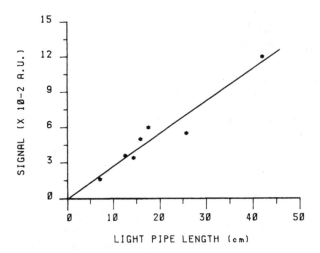

Figure 2-7   Absorbance of ethyl acetate at 1240 cm^{-1} measured in 0.5 ml volume light pipes of varying length.   The data for this graph were obtained from Ref. 19.

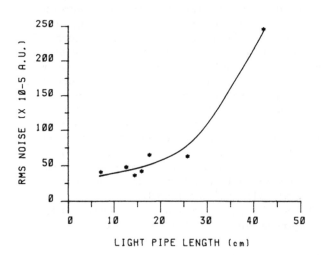

Figure 2-8   Spectral noise measured for 0.5 ml volume light pipes of varying length.   The data for this graph were obtained from Ref. 19.

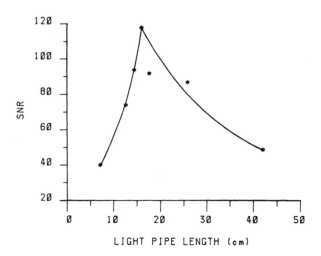

Figure 2-9   Plot of SNR for the 1240 cm^{-1} band of ethyl acetate vs length for 0.5 ml volume light pipes. (Reprinted with permission from Ref. 19.   Copyright © 1984 Society for Applied Spectroscopy, Frederick, Maryland.)

where d_p is the diameter of the light pipe in millimeters, d_f the diameter of the infrared beam at the light pipe entrance in millimeters, and L the length of the light pipe in centimeters. The 0.9 factor in Eq. (2.5) accounts for reflection losses from alkali halide windows attached at each end of the light pipe. The d_p^2/d_f^2 ratio accounts for vignetting losses when the diameter of the light pipe is less than the diameter of the focused infrared beam.   The last term in Eq. (2.5) is an empirical scaling factor that adjusts calculated transmittance values to coincide with an experimentally measured value for a 3 mm ID, 50 cm long light pipe. Because light pipe transmittance (T_P) is inversely proportional to noise and light pipe length (L) is proportional to signal, a plot of LT_P vs L represents the variation of SNR as a function of light pipe length.   The theoretical dependence of light pipe SNR on length for a 0.5 ml light pipe is illustrated in Figure 2-10.   Note the similarity of the shape of this curve with the experimentally derived graph in Figure 2-9.   Figure 2-11 is a family of LT_P vs L curves for light pipe volumes ranging from 0.1 to 4 ml.   The light pipe length that will yield

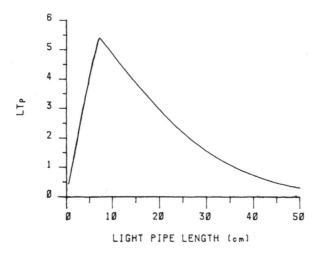

Figure 2-10   Theoretical dependence of SNR (LTp) on light pipe length for a 0.5 ml light pipe.

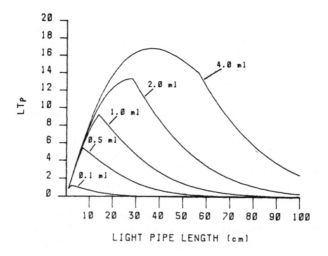

Figure 2-11   LTp vs length curves for light pipes of varying volumes.

the highest SNR depends on its volume. The theoretical optimum length for a 1 ml volume light pipe is about 15 cm, whereas the optimum length for a 4 ml volume is 40 cm. Note that peak LTp values increase with increasing light pipe volume. This behavior would suggest that large-volume light pipes yield the highest sensitivity. However, this trend is observed only when chromatographic elution volumes are equal to or larger than the light pipe volume considered. For example, SNR for a 4 ml elution (FWHM) detected by a 4 ml light pipe would exceed that for a 0.5 ml light pipe because more analyte would be in the infrared beam at peak maximum. However, the spectral SNR of a 0.5 ml elution (FWHM) in a 0.5 ml light pipe should exceed that obtained by using a 4 ml light pipe because the eluent would be diluted by a factor of 8 in the 4 ml light pipe. The curves in Figure 2-11 become broader with increasing light pipe volume. Thus, light pipe length optimization is most critical for small-volume light pipes. For packed column GC separations, light pipes as long as 50 cm have been employed with volumes ranging from 1 to 4 ml. For capillary GC applications, light pipes with 1 mm diameters and 12–15 cm lengths are commonly employed. Capillary GC /FT-IR light pipe volumes are nominally 100 µl.

B. Interface Construction

1. *Light Pipe Coating Procedures*

The development of high-resolution capillary gas chromatography created a need for fabricating highly reflecting, low-volume light pipes. A simple method for making light pipes by coating the inside of small bore glass or quartz tubes with elemental gold was described by Azarraga in 1980 [21]. To assure consistent light pipe dimensions, precision bore tubing was initially recommended for fabricating light pipes. It was later discovered that light pipes made from drawn tubing had smoother internal surfaces than those made from precision bore tubing [22]. Evidently, the boring process creates imperfections in the glass that are not removed by cleaning procedures.

Glass tubing used to fabricate light pipes must be scrupulously clean, straight, and have a constant internal diameter. Several cleaning procedures are recommended prior to coating to ensure high-quality light pipes. Tubing should be presoaked in organic solvent to remove water insoluble oils and then dried by flushing with filtered, dry nitrogen. After overnight soaking in chromic acid to remove oxidizable substances, tubes should be thoroughly rinsed with distilled water. Next, 20 ml of 10% hydrofluoric acid should be poured through the tubes to polish internal surfaces. About 15–20 ml of freshly prepared 0.1 mM stannous chloride solution can be passed through tubes to remove reducible species. After a final rinse with distilled water, tubes should be dried for 4 to 5 hr by passing filtered dry nitrogen through them. After drying, the internal surface of a tube can be coated by dropwise addition of type "N" Hanovia Liquid Bright Gold (Engelhard Industries, East Newark, New Jersey). The liquid gold solution is hydrophobic and great care should be taken to eliminate all traces of water prior to coating. A plug of liquid gold solution is allowed to pass through the tube and excess solution is removed from the bottom of the tube by absorption on tissue paper. After coating, the tube should be dried prior to firing by passing filtered dry nitrogen through it for about 10 hr.

A coated tube can be fired by placing the glass tubing in an oven set at 550°C. Both zone heater ovens [21,22] and annealing ovens [23] have been employed for firing. In both cases, dry air is passed through the tube during firing to burn off solvent and leave behind elemental gold. The reported advantage of using an annealing oven for light pipe firing is that the glass tube can be fired and immediately annealed. Annealed light pipes are less susceptible to coating flaking than tubes that are fired but not annealed [22].

2. *Interface Connections*

Light pipe ends must be sealed in order to conduct GC carrier gas flow. Typically, alkali halide windows are attached to the ends of a light pipe to achieve a seal and permit passage of infrared radiation. Windows can be attached mechanically or cemented to the glass tubing with high-temperature epoxy. GC effluent is directed to the light pipe by means of heated transfer tubing. Transfer tubing can be attached to the light pipe with metal fittings, cemented permanently along with the window, or inserted through small holes in the walls of the light pipe (Figure 2-12). Metal fitting connections [Figure 2-12(a)] have larger dead volumes than other coupling methods and therefore are not preferred. Often GC effluent transfer is achieved by using short lengths of uncoated fused silica capillary tubing [Figure 2-12(b) and 2-12(c)]. When fused silica capillary columns are employed for separation, transfer tubing can be

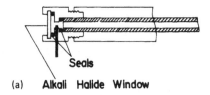

Seals

(a)   **Alkali Halide Window**

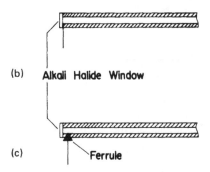

(b)   **Alkali Halide Window**

(c)   **Ferrule**

Figure 2-12   Methods for attaching transfer tubing to a light
pipe. (a) Metal fitting, (b) cemented fused silica tubing, and
(c) ferrule seal.

eliminated and the GC column connected directly to the light
pipe.   Because transfer tubing adds to dead volume mixing,
direct couple connections are always preferable.

3.   Interface Temperature

Gas chromatographic eluent emerges from the chromatographic
column in a vapor state.   To prevent analyte condensation,
the transfer tubing and light pipe must be heated.   Transfer
tubing is often placed inside copper tubing that is heated with
heating tape.   Light pipe and transfer tubing temperatures
should be maintained at least 10°C higher than the highest
column temperature employed during separation.   Light pipe
oven temperature fluctuations should be kept within ±1°C in
order to minimize spectral artifacts produced by black-body
emission.   As shown in Figure 2-13, this effect can be

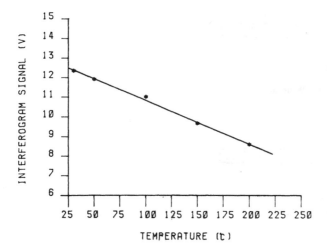

Figure 2-13   Variation of interferogram signal with light pipe oven temperature.

significant for temperatures normally employed in GC/FT-IR. Methods for minimizing this effect are discussed in the next section.

C.   GC/FT-IR Optics

1.   *Single-Beam Configurations*

A variety of GC/FT-IR light pipe optical configurations have been developed over the past 20 years.   Typically, off-axis paraboloid mirrors are used to focus infrared radiation on the light pipe entrance.   Radiation emerging from the light pipe can be collected and focused on a detector by using two paraboloid mirrors (Figure 2-14) or a single ellipsoid (Figure 2-15). Recently, an inexpensive modular light pipe interface was described that was installed within the sample compartment of a commercially available FT-IR [23] (Figure 2-16).   In this design, the light pipe replaces the sample cell holder and the original 190.5 mm EFL 90° off-axis paraboloid mirrors were replaced by 66 mm EFL 90° off-axis paraboloids.   The light pipe was heated by passing current through nichrome wire that was wrapped around it (Figure 2-17).   Detection limits for this light pipe

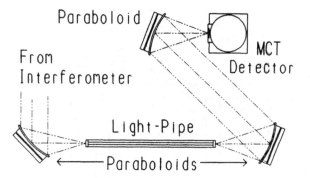

Figure 2-14   GC/FT-IR optical configuration employing off-axis paraboloid mirrors in light pipe collection optics.

interface were shown to be limited by the FT-IR (source, interferometer, detector, electronics, etc.) and not by the light pipe and associated optics.

GC/FT-IR sensitivity at elevated temperatures can be improved by removing unmodulated infrared radiation generated by the light pipe oven [24–27]. Unmodulated radiation provides no useful information but does contribute to detector

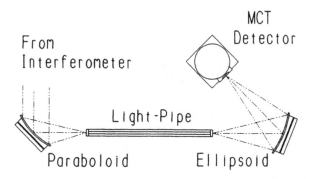

Figure 2-15   GC/FT-IR optical configuration employing an ellipsoid mirror for collecting transmitted radiation.

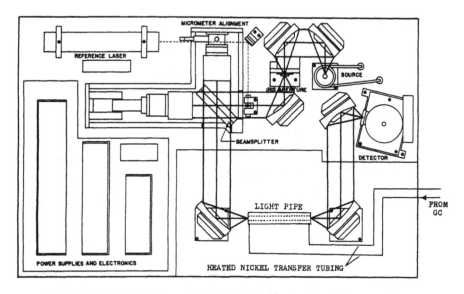

Figure 2-16 Modular GC/FT-IR light pipe interface. (Reprinted with permission from Ref. 23. Copyright © 1987 Society for Applied Spectroscopy, Frederick, Maryland.)

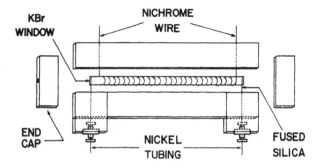

Figure 2-17 Modular GC/FT-IR light pipe interface oven. (Reprinted with permission from Ref. 23. Copyright © 1987 Society for Applied Spectroscopy, Frederick, Maryland.)

saturation. Yang and Griffiths employed beam profiling to characterize the angular distribution of modulated radiation emerging from a light pipe [24]. They concluded that short-focal-length mirrors placed at the light pipe exit lead to detector saturation by transmitting a significant portion of unmodulated infrared radiation to the detector. This effect can be minimized by employing long-focal-length mirrors (f/5 or f/6) for light pipe collecting optics to discriminate between modulated and unmodulated radiation (Figure 2-18). Brown et al. attached a water-cooled extension to the beam exit of a light pipe interface and effectively reduced unmodulated radiation emanating from the end of the heated light pipe [25]. Buijs et al. described yet another method of selectively eliminating unmodulated infrared radiation emitted by the light pipe oven. They focused radiation from the light pipe exit onto a variable aperture that was adjusted to pass only the portion of the light pipe image comprising modulated radiation [26] (Figure 2-19).

The image size formed by light pipe collecting optics determines the detector area that should be employed for optimum sensitivity. MCT detectors ranging in diameter from 50 μm to 2 mm have been employed for GC /FT-IR. Henry et al. studied the effect of detector size on GC /FT-IR performance and concluded that a 0.5 mm diameter MCT detector was optimum for capillary GC /FT-IR applications [27]. Detectors of this size are readily available from commercial sources.

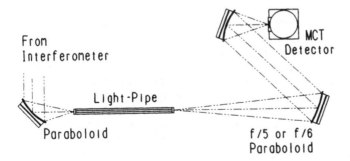

Figure 2-18 Long-focal-length collection optics designed to minimize detection of light pipe oven black-body emission.

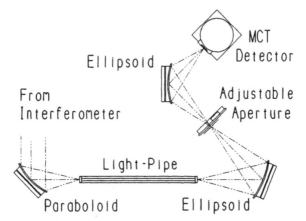

Figure 2-19   Optical configuration for a GC/FT-IR light pipe
interface incorporating an adjustable aperture to discriminate
between modulated and unmodulated infrared radiation.

2.   Double-Beam Interferometry

One of the first attempts to combine gas chromatography with
infrared detection employed double-beam interferometry for
eluent detection [12].   The concept of double-beam interferometry was first proposed by Fellgett in 1958 [28].   Double-beam
interferometry provides enhanced sensitivity when spectral
noise would otherwise be digitization-noise-limited [29,30].   A
double-beam GC/FT-IR interface incorporating cube corner
mirrors in the interferometer is depicted in Figure 2-20.

Michelson interferometers generate two modulated beams that
are 180° out of phase with each other.   Single-beam FT-IR
instruments collect only one of these beams because the second
beam returns to the infrared source.   With appropriate optics
(Figure 2-20), it is possible to access both interferometer
beams and combine these beams optically or electronically.
Beam recombination produces a nulled interferogram with a significantly lower dynamic range than single-beam interferograms
(Figure 2-21).   Decreased interferogram dynamic range permits
more precise signal digitization in areas away from zero path
difference.   Information lost by nulling comprises background
spectral features that are common to both beam paths.   These
common features are eliminated in conventional FT-IR by

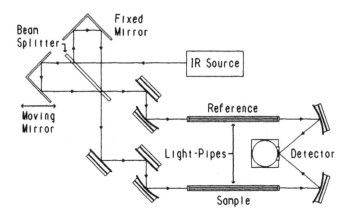

Figure 2-20   Optical configuration of a double-beam GC /FT-IR light pipe interface.

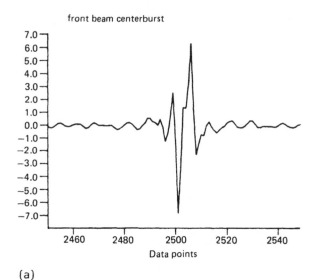

(a)

Figure 2-21   Demonstration of double-beam interferometry nulling. (a) Front-beam path interferogram signal, (b) back-beam path signal, and (c) measured sum of both beam path signals.

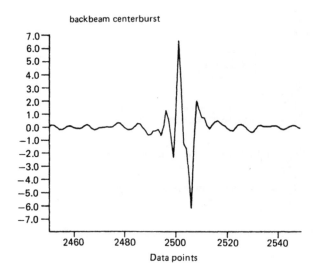

(b)

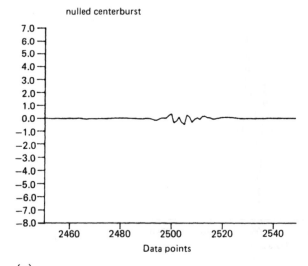

(c)

Figure 2-21 (Continued)

ratioing single-beam spectra derived from sample and reference measurements. Double-beam GC/FT-IR has been shown to provide significant sensitivity improvements compared with single-beam detection when high-transmittance light pipes are employed [15,16]. For capillary GC/FT-IR applications, double-beam FT-IR provides little improvement. The dynamic range of the interferogram signal is effectively reduced by the low transmittance of small-aperture (1 mm diameter) light pipes typically employed.

IV. MATRIX ISOLATION

Whittle, Dows, and Pimentel reported the first matrix isolation infrared spectrum in 1954 [31]. Subsequently, matrix isolation vibrational spectroscopy became an important tool for investigating properties of transient species. The matrix isolation technique has been incorporated in the design of a sensitive interface for GC/FT-IR (GC/MI/FT-IR) [10,13,32–40]. GC/MI/FT-IR interface designs have evolved rapidly since the introduction of the technique in 1979. GC/MI/FT-IR became commercially available in 1984 from Mattson Instruments, Inc. under the name Cryolect^{TM}. The basic design of the Cryolect^{TM} is shown in Figure 2-22. Gas chromatographic eluent and argon matrix gas are co-deposited at 10°K on the edge of a 4 in. diameter gold-coated collection disk. The collection disk is slowly rotated during gas chromatographic separation and eluent is deposited as a thin band on the surface of the disk. Five hours of chromatography can be frozen on the disk at one time. Infrared spectra are measured after gas chromatographic separation has finished. An FT-IR infrared beam is reflected through deposited eluent and focused on an MCT detector (Figure 2-23). The most recent version of the Cryolect^{TM} incorporates a vacuum FT-IR interferometer with 1 in. optics (Figure 2-24). Nitrogen purge is not required for this system because the entire beam path is contained within vacuum. Subnanogram detection limits can be routinely attained by GC/MI/FT-IR.

A. Interface Considerations

1. *Deposition Rate*

Cryogenic temperatures must be maintained during GC eluent trapping in order to preserve chromatographic resolution and

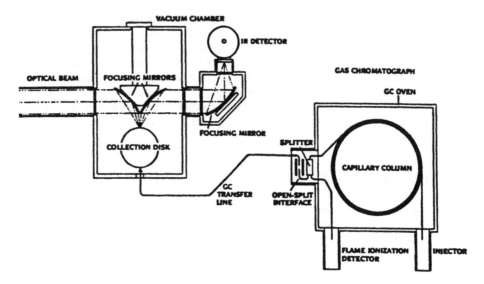

Figure 2-22   Schematic of the Cryolect^{TM} GC/MI/FT-IR interface. (Reprinted from Ref. 36.   Copyright © 1984 by International Scientific Communications, Inc., Fairfield, Connecticut.)

ensure that eluent species do not aggregate on the collection
disk. As matrix gas and eluent condense, energy is released
in the form of the heat of fusion for these species. At high
deposition rates, this heat may increase the temperature of the
collection disk substantially, permitting isolated species migration.   This would degrade chromatographic resolution and permit eluents with high affinity to associate. Eluent aggregation
would produce infrared spectral artifacts, making it more difficult to identify species present. This problem can be avoided
by controlling the rate of matrix deposition. The heat load (Q)
generated by a condensing matrix is given by [32]

$$Q = kAt^{-1} \Delta T \tag{2.6}$$

where k is the thermal conductivity of the matrix, A the area
of the matrix, t the matrix thickness, and ΔT the temperature
gradient across the matrix. For a 0.5 mm thick, 10 mm^2 argon
matrix formed at a rate of 4×10^{-5} mol/min (1 ml/min), the

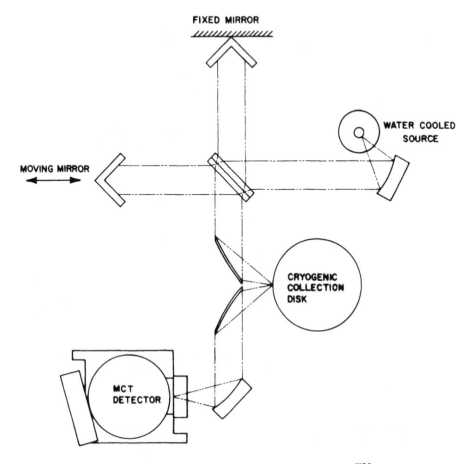

Figure 2-23   Optical configuration of the Cyrolect^{TM} interface.
(Courtesy of Mattson Instruments, Inc., Madison, Wisconsin.)

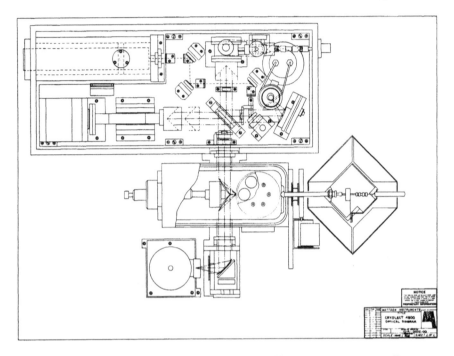

Figure 2-24   Schematic of the Cryolect^{TM} 4800 GC/MI/FT-IR interface. (Courtesy of Mattson Instruments, Inc., Madison, Wisconsin.)

temperature gradient (ΔT) is 1°K. These conditions are accep-table for matrix isolation and are typical of GC/MI/FT-IR.

2. *Interface Parameters*

Under normal separating conditions, the average concentration of eluents in GC carrier gas is about 1 part in 10,000. If carrier gas were used as the matrix gas, matrix species would exceed eluent by a factor of 10^4 and the enrichment factor for the interface would be unity. In practice, matrix/eluent ratios of 100:1 are adequate for minimizing eluent-eluent interactions. A 100:1 matrix/eluent ratio would yield an enrichment factor of about 100. In order to achieve matrix/eluent ratios of 100:1, 99% of the carrier gas must be selectively removed prior to matrix deposition. This is achieved by using a helium (99%) and argon (1%) mixture for GC carrier gas. Helium is not condensed at 10°K and is removed by the vacuum system. The argon is condensed and serves as the matrix for eluent isolation.

Commercially available GC/MI/FT-IR systems are limited by the infrared beam diameter to infrared spectral analysis at 6 sec chromatographic intervals. For intermediate bore capillary column separations, average GC elution volumes are about 100 μl. For a typical capillary column flow rate of 1 ml/min, the duration of an average GC elution would be about 10 sec. If the elution maximum were detected at some position on the collection disk, eluent spectra would also appear for measurements immediately preceding and following that measurement, resulting in a detected elution width of 18 sec. Thus, resolution degradation (R_d) in this case would be

$$R_d = \frac{R_{CHROM}}{R_{MI}} = \frac{\sigma_{MI}}{\sigma_{CHROM}} = \frac{4.5s}{2.5s} = 1.8 \tag{2.7}$$

For narrow bore capillary column GC separations, σ_{CHROM} would be smaller and R_d would therefore be larger.

All of the eluent from the GC column could, in principle, be trapped on the matrix isolation collection disk. In practice, some of the GC effluent is split to a second detector. It is common for 5–20% of the column effluent to be diverted. Flame ionization detectors (FID) and mass spectrometers (MS) have both been employed as second detectors. These detectors are more sensitive than FT-IR and therefore do not require an equal portion of GC eluent in order to attain comparable detection limits. The GC/MI/FT-IR/MS combination is particularly useful as it provides complementary structure-specific infrared and mass spectra for each eluent.

Applying Beer's law to the special case of matrix isolation trapping yields the following equation [32]:

$$A = a_{MI}Q_A/S_M \tag{2.8}$$

where A is absorbance, a_{MI} the eluent absorptivity, Q_A the quantity of eluent trapped, and S_M the cross-sectional area of the matrix. Smaller matrix deposition areas (S_M) yield larger absorbances. Thus, eluent should be confined to the smallest possible area on the collection disk. Typically, a matrix deposition band that is less than 0.5 mm in diameter is employed. Note that eluent absorptivity (a_{MI}) denoted in Eq. (2.8) is *not* equal to vapor phase absorptivity (a_{VP}). Rotations of

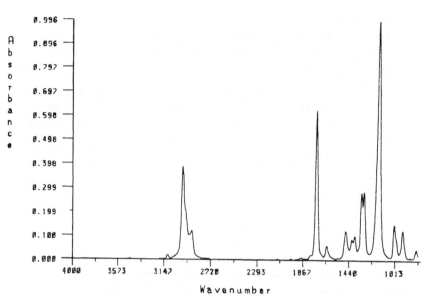

(a)

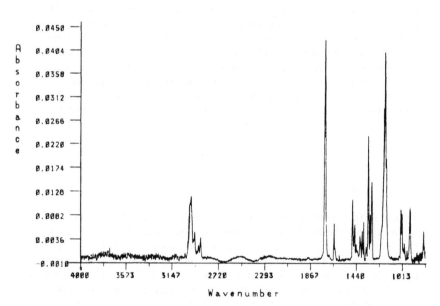

(b)

Figure 2-25   Comparison of (a) vapor phase and (b) matrix-isolated infrared spectra for isobutyl methacrylate.

matrix-isolated species are hindered, whereas vapor phase species are free to rotate. As a result, matrix isolation bands are sharper and more intense than vapor phase absorptions ($a_{MI} > a_{VP}$).

Larger enrichment factors and higher eluent absorptivity account for much of the sensitivity improvement of GC /MI /FT-IR compared to light pipe GC /FT-IR. In addition to these factors, spectral resolution and signal averaging are unrestricted for GC /MI /FT-IR measurements because measurements are made after chromatographic separation. In contrast, light pipe GC / FT-IR measurements are restricted to low-resolution and signal-averaging periods are limited to the time required for chromatographic elution. Enrichment, increased absorptivity for matrix-isolated species, and signal-averaging advantages combine to yield a 100-fold lower GC /MI /FT-IR limit of detection relative to light pipe interfaces (Figure 2-2).

B. Data-Acquisition Requirements

As mentioned previously, matrix isolation infrared spectra are characterized by sharp absorption bands. Sharp bands result from the combined effects of minimal intermolecular interaction and the absence of rotational band broadening. Coleman and Gordon have compared spectral features of matrix-isolated species with corresponding vapor phase spectra for a wide range of organic functionalities [41–51]. Removal of rotational broadening reveals a wealth of vibrational structure in matrix isolation spectra that would otherwise be hidden. This is illustrated for isobutyl methacrylate in Figure 2-25. A number of vibrational absorptions can be discerned in the matrix isolation spectrum that are not resolved in the vapor phase spectrum. Some matrix isolation vibrational features have bandwidths of 1 cm^{-1} or less, requiring the use of FT-IR spectrometers with moderate wavelength resolution for accurate representation. GC /MI /FT-IR spectra are usually measured at 1 cm^{-1} resolution in order to preserve these features.

V. SUBAMBIENT TRAPPING

The technique of subambient trapping GC /FT-IR takes advantage of the eluent enrichment feature of matrix isolation without achieving molecular isolation. Shafer et al. demonstrated that low-volatility gas chromatographic eluents could be deposited on

infrared transparent substrates at ambient temperature and
subsequently analyzed by infrared microscopy [11]. This concept was extended to the analysis of higher volatility substances by cooling the infrared substrate to subambient temperatures [52]. The trapping surface was a zinc selenide
window that was maintained at subambient temperature by using a thermoelectric (Peltier) cooler. Gas chromatographic
eluent passed through a heated 50 μm ID fused silica restrictor
and was deposited on the cooled window (Figure 2-26). FT-IR
microscopy [53] was employed to obtain infrared spectra of
trapped eluents. The window was mounted on a movable stage
that was placed into the beam path of a commercial FT-IR
microscopy accessory for analysis and chromatogram generation. Detection limits that are competitive with matrix isolation
have been reported for this interface.

A. Trapping Efficiency

The trapping efficiency for eluents deposited on the infrared-
transmitting window depends on the positioning of the restrictor, temperature of the window, and volatility of the eluent.
The optimum distance between the restrictor and window surface was empirically found to be equal to the diameter of the
restrictor. Thus, a 50 μm ID restrictor would be placed approximately 50 μm from the deposition surface. If the restrictor is
closer to the surface, convective cooling can clog the restrictor.
If the restrictor is farther from the surface, eluent may be
deposited over larger areas on the window, yielding higher
detection limits. Unlike matrix isolation, subambient trapping
temperatures employed thus far have not been low enough to
prevent evaporation of volatile eluents. Figure 2-27 shows the
effect of window temperature on detected absorbance for nitrobenzene depositions. A temperature of $-30°C$ was required in
order to prevent rapid evaporation of deposited eluent.

The temperature of the deposition window can also determine
the physical state of deposited eluent. For example, the spectrum of nitrobenzene measured at $-45°C$ has a sloping baseline
caused by scattering, indicating that deposited eluent was
crystalline [Figure 2-28(a)]. In contrast, the spectrum measured
at $-30°C$ exhibits no scattering features, indicating that the
eluent may be in a liquid state rather than crystalline [Figure
2-28(b)]. For nonvolatile components, 100% yield can easily be
achieved for subambient trapping interfaces. Also, enrichment

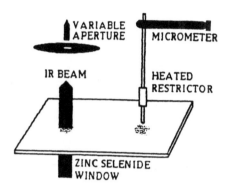

Figure 2-26   Diagram of a subambient trapping GC/FT-IR
interface. (Reprinted with permission from Ref. 14. Copyright
© 1987, Pergamon Press, New York.)

factors for this interface are large because all chromatographic
mobile phase is removed during eluent deposition.

B. Detection Requirements

Subambient GC/FT-IR interface designs borrow much from GC/
MI/FT-IR technology. Instrumental differences between the
two interfaces reflect deposition surface temperature requirements. GC/MI/FT-IR requires 10°K temperatures in order to
prevent eluent migration and maintain a rigid argon cage.
Subambient GC/FT-IR temperatures need only be low enough to
permit infrared analysis before eluents evaporate from the
surface. Subambient GC/FT-IR infrared spectra have been
measured by means of infrared microscopy. This method of
analysis was employed more for convenience than necessity.
An optical configuration similar to that employed for GC/MI/
FT-IR (Figure 2-23) should be compatible with subambient trapping techniques. The optical configuration for trapping interfaces determines the resolution degradation of the interface.
Therefore, resolution degradation for subambient trapping GC/
FT-IR should be about the same as for GC/MI/FT-IR. A subambient trapping system named "Tracer" is now commercially
available from Digilab, Inc., Cambridge, Massachusetts.

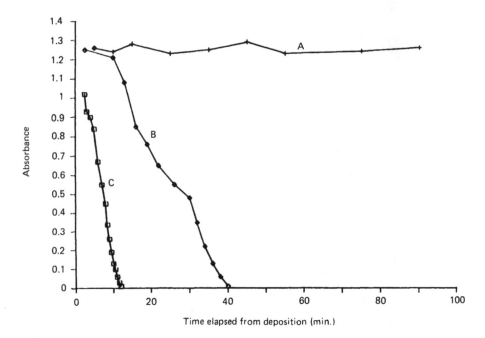

Figure 2-27   Effect of window temperature on quantity of nitrobenzene remaining on the surface of the window.   Nitrobenzene
absorbance at 1524 cm^{-1} was monitored for 150 ng depositions.
(a) -30°C, (b) -20°C, and (c) -10°C.   (Reprinted with permission from Ref. 52.   Copyright © 1986 American Chemical
Society, Washington, D.C.)

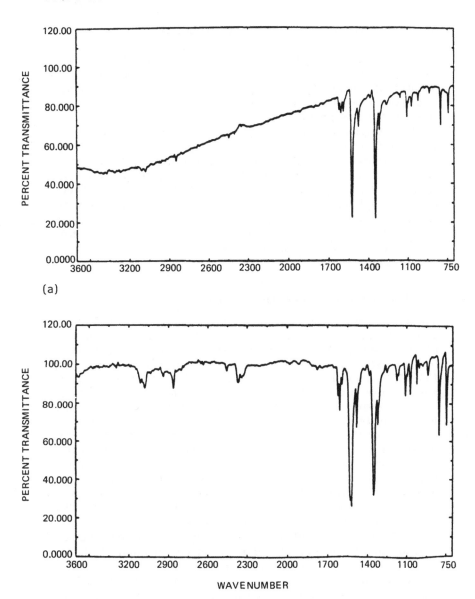

(a)

(b)

WAVENUMBER

Figure 2-28 Subambient trapping GC /FT-IR spectra for nitro-benzene at (a) −45°C and (b) −30°C. (Reprinted with permission from Ref. 52. Copyright © 1986 American Chemical Society, Washington, D.C.)

VI. AUGMENTED GC/FT-IR INSTRUMENTATION

The power of GC/FT-IR analysis can be significantly increased
by modifications to the basic interface. The component-
separating ability of gas chromatography can be enhanced by
employing two-dimensional ("heart-cut") chromatography.
Heart cutting is particularly useful for the analysis of extremely complex mixtures that may not be completely resolved
by a single column.

Qualitative structural information provided by FT-IR analysis can be augmented by the addition of complementary detectors. A judicious choice for a second detector is mass spectrometry. Complementary information provided by simultaneous infrared and mass spectroscopic analysis may permit unequivocal
component identification in cases for which ambiguous identifications would be obtained by infrared or mass spectroscopic
measurements alone.

A. Two-Dimensional Capillary GC/FT-IR

The use of two chromatographic columns containing different
stationary phases to enhance GC/FT-IR analysis of complex
mixture components was first reported by Azarraga and Potter
in 1981 [54]. This concept was subsequently refined by Smith
[55—57]. Two-dimensional capillary GC/FT-IR instrumentation
incorporates two capillary columns contained in a dual oven gas
chromatograph (Figure 2-29). A micro switching valve is used
to selectively divert portions of gas chromatographic eluent
from one of these columns to the other. In this manner, components that co-elute from the first column can be separated by
the second. Separations on the second column are not complicated by the wide range of original mixture components because
only a small number of chromatographic eluents from the first
column are transported to the second. The procedure is illustrated in Figure 2-30. The chromatogram in Figure 2-30(a) was
obtained by using a DB-5 capillary column. Chromatographically
unresolved components labeled as heart-cut in Figure 2-30(a)
were diverted from the DB-5 column into a DB-WAX column.
The chromatogram obtained by using the DB-WAX column is
shown in Figure 2-30(b). The unresolved chromatographic peak
in Figure 2-30(a) was separated into five major components and
several minor components [Figure 2-30(b)]. Infrared spectra
for the major components are shown in Figure 2-31. Spectra in
Figure 2-31 represent a variety of compound types that were

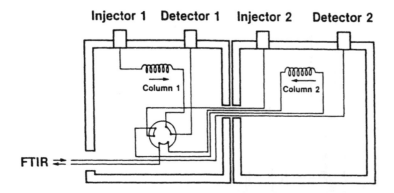

Injector 1   Detector 1   Injector 2   Detector 2

FTIR ⇌

Figure 2-29   Diagram of a dual oven GC/FT-IR analysis system.   (Reprinted from Ref. 57.   Copyright © 1986 Society for Applied Spectroscopy, Frederick, Maryland.)

not separated by the first column but readily separated by the second.

B.   Integrated GC/FT-IR/MS

The concept of *hyphenated* analysis techniques has been discussed in detail by Hirschfeld [58].   Combining information derived from complementary analysis methods can increase confidence in analytical determinations.   The most common structure-specific detectors employed with gas chromatography are infrared and mass spectrometry.   In most instances, GC/MS or GC/FT-IR analysis alone does not provide unequivocal mixture-component identification.   Infrared spectroscopy can provide functional group determinations but often cannot distinguish between members of a homologous series.   Mass spectrometry provides molecular weight information and can distinguish homologues.   However, isomer differentiation by mass spectrometry is difficult.   The complementary nature of infrared and mass analysis is illustrated by Figures 2-32 to 2-35.   Figure 2-32 contains mass spectra for the three xylene isomers and illustrates the failure of mass spectrometry to distinguish isomeric species.   Mass spectra shown in Figure 2-32 are virtually identical.   In contrast, infrared spectra for the xylenes contain large differences in the fingerprint region permitting distinction

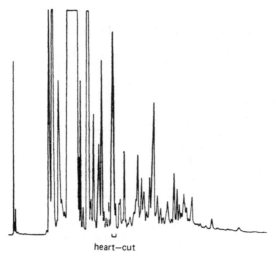

heart—cut

(a)

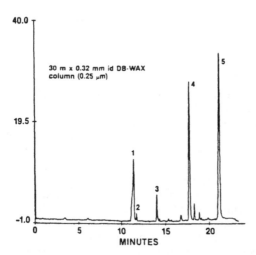

(b)

Figure 2-30   (a) Flame ionization chromatogram for a terpene mixture. The portion labeled heart-cut was diverted to a DB-WAX capillary column.   (b) Gram—Schmidt GC/FT-IR chromatogram for heart-cut portion indicated in (a). (Reprinted from Ref. 57. Copyright © 1986 Society for Applied Spectroscopy, Frederick, Maryland.)

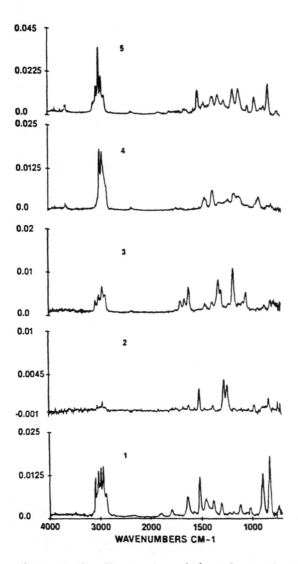

Figure 2-31   Vapor phase infrared absorbance spectra obtained
from the chromatographic data shown in Figure 2-30(b).   Spectrum numbers correspond to chromatographic peaks indicated in
Figure 2-30(b).   (Reprinted from Ref. 57.   Copyright © 1986
Society for Applied Spectroscopy, Frederick, Maryland.)

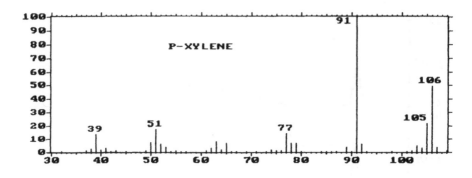

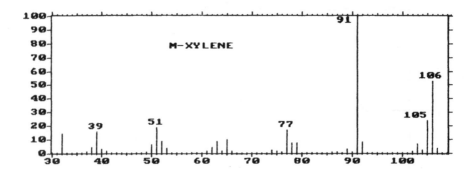

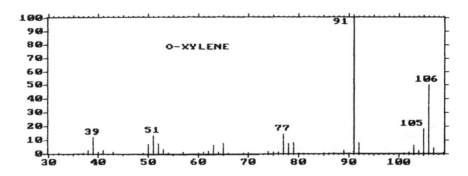

Figure 2-32   Mass spectra of the xylene isomers.

P-XYLENE

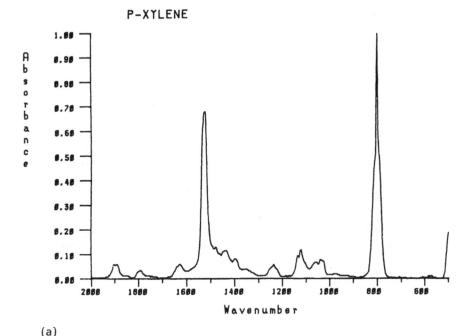

(a)

Figure 2-33   Infrared spectra of the (a) p-, (b) m-, and (c) o-xylene isomers.

(Figure 2-33).   Figure 2-34 contains infrared spectra for octane, decane, and dodecane.   Infrared spectra for these alkanes are nearly identical, indicating a lack of homologue discrimination by infrared spectrometry.   However, mass spectra for these species are readily distinguished (Figure 2-35).   Complementary infrared and mass spectrometry have been combined for the analysis of mixture components eluting from a single-gas chromatograph [59–66].

Wilkins et al. first demonstrated unambiguous qualitative analysis for mixture components by GC /FT-IR /MS in 1981 [59].

M-XYLENE

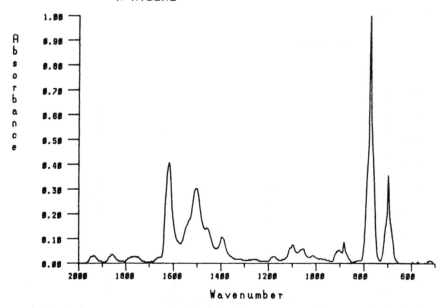

(b)

O-XYLENE

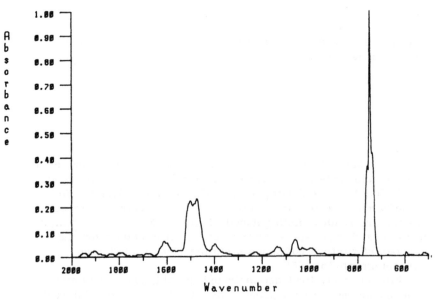

oxylene.abs

(c)

Figure 2-33   (Continued)

n-OCTANE

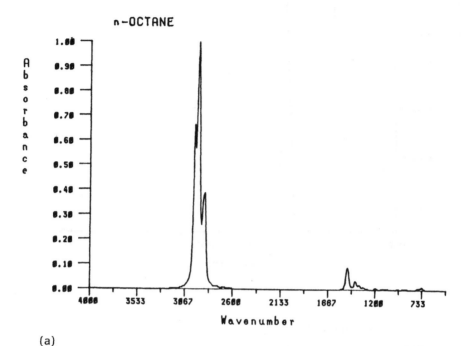

(a)

Figure 2-34   Infrared spectra of (a) *n*-octane, (b) *n*-decane, and (c) *n*-dodecane.

The integrated GC /FT-IR /MS system used for their measurements is illustrated in Figure 2-36. Approximately 1% of the gas chromatographic eluent was diverted to the mass spectrometer. The high split ratio (99:1) was needed to compensate for differences in FT-IR and mass spectrometer sensitivities. By using complementary information derived from the combined analysis methods, 13 of 17 components of known mixtures were correctly and unambiguously identified (Table 2-1).

In 1982, Crawford et al. described a GC /FT-IR /MS analysis system based on capillary column (SCOT) separations [60]. Capillary GC /FT-IR /MS performance was evaluated for both serial and parallel FT-IR /MS connections [61]. It was determined that the light pipe dead volume degraded chromatographic resolution excessively when detectors were connected in serial. A low-cost GC /FT-IR /MS system has recently been described

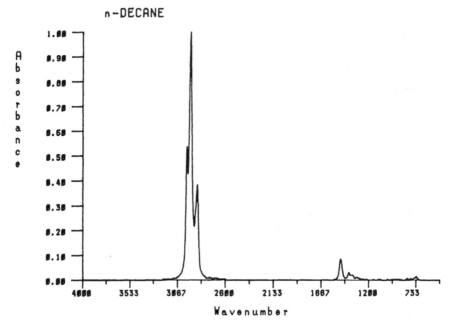

(b)

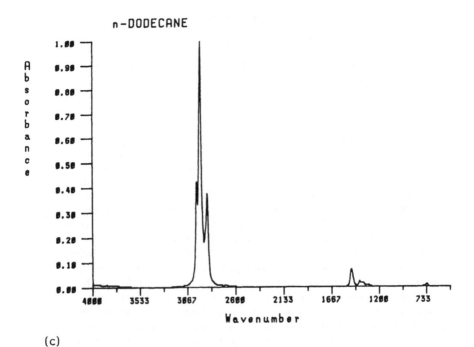

(c)

Figure 2-34   (Continued)

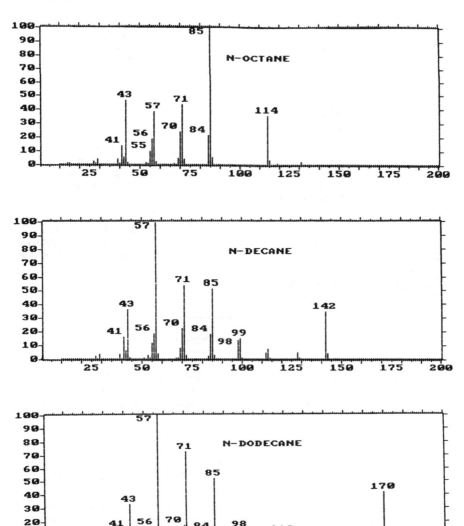

Figure 2-35   Mass spectra of *n*-octane, *n*-decane, and
n-dodecane.

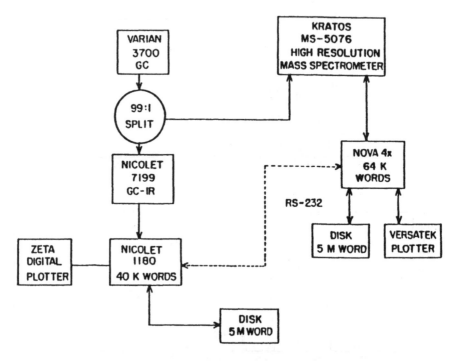

Figure 2-36   Block diagram for an integrated GC/FT-IR/MS
analysis system.   (Reprinted with permission from Ref. 59.
Copyright © 1981 American Chemical Society, Washington, D.C.)

that employs a low-resolution mass selective detector (MSD) [65].
GC/FT-IR/MS analysis systems are commercially available from
at least two instrument manufacturers.

The efficacy of GC/FT-IR/MS complex mixture analysis is
heavily dependent on the method selected for interpreting infrared and mass spectrometric information obtained during chromatographic separation.   Initial GC/FT-IR/MS identifications were
performed by manually comparing library search results for
infrared and mass spectrometric data [59,60].   Williams et al.
proposed several methods for automated GC/FT-IR/MS data
evaluation [67].   Infrared information could be used as a functional group classifier or prefilter prior to mass spectral

Table 2-1   Infrared and Mass Spectrometry Search Results[a]

Sample	Components	IR	MS	Combined
A	*p*-xylene	*	**	+
	m-xylene	*	**	+
	o-xylene	*	**	+
B	anisole	−	4	N
	acetophenone	*	2	+
	methyl *o*-toluate	*	*	+
	methyl salicylate	*	4	+
C	benzene	*	5	+
	toluene	*	5	+
	p-xylene	*	**	+
D	3-methyl-cyclopentene	−	−	N
	cyclohexene	*	2	+
E	*trans, trans*-2,4-hexadiene	**	**	N
	cis, trans-2,4-hexadiene	**	**	N
F	*n*-pentane	**	4	+
	n-hexane	**	3	+
	n-heptane	**	2	+

[a]One asterisk indicates compound was among the five best search matches; double asterisk indicates that all mixture components were among the five best search matches; minus sign indicates compound was not among the five best search matches; a number indicates the number of duplicate spectra found among the best five search matches; N indicates ambiguous or erroneous identification; plus sign indicates correct identification.

Source: Reprinted with permission from Ref. 59. Copyright © 1981 American Chemical Society, Washington, D.C.

Table 2-2   Processing Times for Integrated GC /FT-IR /MS
Identifications

Procedure	Computers used	Time (min)
Integrated GC /FT-IR /MS run	two Nicolet 1280s	25
MS and IR Fourier transforms	two Nicolet 1280s	60
MS search	Vax 11/750	30
IR search	Nicolet 1280	
Search comparison	Nicolet 1280	5
Molecular ion algorithm	Vax 11/750-Nicolet 1280	20
Total time		140

Source: Reprinted with permission from Ref. 63. Copyright ©
1984 American Chemical Society, Washington, D.C.

searching. Alternatively, a combination library consisting of
mass and infrared spectra of reference materials could be
created and used for searching [68]. The major drawback of
this method is that the combined library would be limited to
the small size of available vapor phase infrared libraries. The
largest vapor phase libraries contain approximately 10,000
spectra, whereas the Wiley /NBS mass spectral database contains
79,560 spectra of 67,128 different compounds [69]. Laude et al.
used a postsearch accurate-mass-measurement filter to improve
compound identifications by GC /FT-IR /MS [70]. By using
Fourier transform mass spectrometry, electron impact and
chemical ionization mass spectra were obtained for eluting components during a single separation [63]. This capability provided additional structural information for GC /FT-IR /MS analyses. Data processing time for a 17-component GC /FT-IR /MS
separation employing electron impact and chemical ionization
mass spectrometry is broken down by function in Table 2-2.

The total analysis time was 140 min. Future efforts in GC /FT-
IR /MS instrumentation development will likely concentrate on
improving data-evaluation algorithms to speed automated analysis.

VII. SUMMARY

A comparison of interface characteristics for the three GC /FT-IR
methodologies is given in Table 2-3. Enhanced sensitivity for
matrix isolation and subambient trapping interfaces derives
from the large enrichment factors achieved by eluent trapping.
All methods degrade chromatographic resolution to some extent.
Light pipe interfaces are less expensive and less sensitive than
trapping techniques but are still useful for many capillary
GC /FT-IR applications. Subambient trapping interfaces show
great promise for future applications. These interfaces do not
require expensive cryogenic cooling systems and take full advantage of eluent enrichment sensitivity enhancement. GC /MI /
FT-IR component specificity is unmatched by any other GC /FT-
IR interface. Matrix isolation absorbance bands are extremely
sharp and highly representative of intramolecular environment.
GC /MI /FT-IR interfaces are expensive, but the cost of these
interfaces should decrease as designs are refined.

Table 2-3   Comparison of GC /FT-IR Interface
Parameters

Method	R_d	Enrichment factor	% Yield
Light pipe	1.2	0.81	100
GC /MI /FT-IR	1.8	100	100
Subambient	1.8	infinite	100

REFERENCES

1. J. U. White, N. L. Alport, W. M. Ward, and W. S. Gallaway, *Anal. Chem.*, *31*: 1267 (1959).

2. J. E. Stewart, R. O. Brace, T. Johns, and W. F. Ulrich, *Nature, 186*: 628 (1960).

3. W. L. Senn and H. V. Drushel, *Anal. Chim. Acta.*, *25*: 328 (1961).

4. J. Haslam, A. R. Jeffs, and H. A. Wilks, *Analyst, 86*: 44 (1961).

5. M. St. C. Flett and J. Hughes, *J. Chromatogr.*, *11*: 434 (1963).

6. A. M. Bartz and H. D. Ruhl, *Anal. Chem.*, *36*: 1892 (1964).

7. P. A. Wilks, Jr. and R. A. Brown, *Anal. Chem.*, *36*: 1896 (1964).

8. T. Tsuda, H. Mori, and D. Ishii, *Bunsehi Kagaku, 18*: 1328 (1969).

9. M. J. D. Low and S. K. Freeman, *Anal. Chem.*, *39*: 194 (1967).

10. S. Bourne, G. T. Reedy, and P. T. Cunningham, *J. Chromatogr. Sci.*, *17*: 460 (1979).

11. K. H. Shafer, P. R. Griffiths, and R. Fuoco, *J. High Res. Chromatogr. Chromatogr. Commun.*, *9*: 124 (1986).

12. M. J. D. Low, *Anal. Lett.*, *1*: 819 (1968).

13. G. T. Reedy, D. G. Ettinger, S. Bourne, and J. F. Schneider, "Abstracts of the 35th Annual Pittsburgh Conference on Analytical Chemistry and Applied Spectroscopy," Atlantic City, New Jersey, March 1984, paper 363.

14. P. R. Griffiths and D. E. Henry, *Prog. Anal. Spectrosc.*, *9*: 455 (1986).

15. M. M. Gomez-Taylor and P. R. Griffiths, *Anal. Chem.*, *50*: 422 (1978).

16. D. Kuehl, G. J. Kemeny, and P. R. Griffiths, *Appl. Spectrosc.*, *34*: 222 (1980).

17. P. J. Coffey, D. R. Mattson, and J. C. Wright, *Am. Lab. (Fairfield, Conn.)*, *10* (5): 126 (1978).

18. D. L. Wall and A. W. Mantz, *Appl. Spectrosc.*, *31*: 552 (1977).

19. G. N. Giss and C. L. Wilkins, *Appl. Spectrosc.*, *38*: 17 (1984).

20. P. R. Griffiths, *Appl. Spectrosc.*, *31*: 284 (1977).

21. L. V. Azarraga, *Appl. Spectrosc.*, *34*: 224 (1980).

22. P. W. J. Yang, E. L. Ethridge, J. L. Lane, and P. R. Griffiths, *Appl. Spectrosc.*, *38*: 813 (1984).

23. R. E. Fields III and R. L. White, *Appl. Spectrosc.*, *41*: 705 (1987).

24. P. W. J. Yang and P. R. Griffiths, *Appl. Spectrosc.*, *38*: 816 (1984).

25. R. S. Brown, J. R. Cooper, and C. L. Wilkins, *Anal. Chem.*, *57*: 2275 (1985).

26. H. Buijs, J. N. Berube, and G. Vail, "Abstracts of the 35th Annual Pittsburgh Conference on Analytical Chemistry and Applied Spectroscopy," Atlantic City, N.J., March 1984, paper 273.

27. D. E. Henry, A. Giorgetti, A. M. Haefner, P. R. Griffiths, and D. F. Gurka, *Anal. Chem.*, *59*: 2356 (1987).

28. P. Fellgett, *J. Phys. Rad.*, *19*: 237 (1958).

29. D. Kuehl, G. J. Kemeny, and P. R. Griffiths, *Appl. Spectrosc.*, *34*: 95 (1980).

30. D. L. Beduhn and R. L. White, *Appl. Spectrosc.*, *40*: 628 (1986).

31. E. Whittle, D. A. Dows, and G. C. Pimentel, *J. Chem. Phys.*, *22*: 1943 (1954).

32. G. T. Reedy, S. Bourne, and P. T. Cunningham, *Anal. Chem.*, *51*: 1535 (1979).

33. D. M. Hembree, A. A. Garrison, R. A. Crocombe, R. A. Yokley, E. L. Wehry, and G. Mamantov, *Anal. Chem.*, *53*: 1783 (1981).

34. G. Mamantov, A. A. Garrison, and E. L. Wehry, *Appl. Spectrosc.*, *36*: 339 (1982).

35. A. A. Garrison, G. Mamantov, and E. L. Wehry, *Appl. Spectrosc.*, *36*: 348 (1982).

36. S. Bourne, G. T. Reedy, P. J. Coffey, and D. Mattson, *Am. Lab. (Fairfield, Conn.)*, *16* (6): 90 (1984).

37. G. T. Reedy, D. G. Ettinger, J. F. Schneider, and S. Bourne, *Anal. Chem.*, *57*: 1602 (1985).

38. J. R. Schneider, G. T. Reedy, and D. G. Ettinger, *J. Chromatogr. Sci.*, *23*: 49 (1985).

39. C. J. Wurrey, S. Bourne, and R. D. Kleopfer, *Anal. Chem.*, *58*: 482 (1986).

40. J. F. Schneider, J. C. Demirgian, and J. C. Stickler, *J. Chromatogr. Sci.*, *24*: 330 (1986).

41. W. M. Coleman III and B. M. Gordon, *Appl. Spectrosc.*, *41*: 886 (1987).

42. W. M. Coleman III and B. M. Gordon, *Appl. Spectrosc.*, *41*: 1159 (1987).

43. W. M. Coleman III and B. M. Gordon, *Appl. Spectrosc.*, *41*: 1163 (1987).

44. W. M. Coleman III and B. M. Gordon, *Appl. Spectrosc.*, *41*: 1169 (1987).

45. W. M. Coleman III and B. M. Gordon, *Appl. Spectrosc.*, *41*: 1431 (1987),

46. W. M. Coleman III and B. M. Gordon, *Appl. Spectrosc.*, *42*: 101 (1988).

47. W. M. Coleman III and B. M. Gordon, *Appl. Spectrosc.*, *42*: 108 (1988).

48. W. M. Coleman III and B. M. Gordon, *Appl. Spectrosc.*, *42*: 304 (1988).

49. W. M. Coleman III and B. M. Gordon, *Appl. Spectrosc.*, *42*: 666 (1988).

50. W. M. Coleman III and B. M. Gordon, *Appl. Spectrosc.*, *42*: 671 (1988).

51. W. M. Coleman III and B. M. Gordon, *Appl. Spectrosc.*, *42*: 1049 (1988).

52. R. Fuoco, K. H. Shafer, and P. R. Griffiths, *Anal. Chem.*, *58*: 3249 (1986).

53. J. E. Katon, G. E. Pacey, and J. F. O'Keefe, *Anal. Chem.*, *58*: 465A (1986).

54. L. V. Azarraga and C. A. Potter, *J. High Res. Chromatogr. Chromatogr. Commun.*, *4*: 60 (1981).

55. S. L. Smith, *J. High Res. Chromatogr. Chromatogr. Commun.*, *8*: 385 (1985).

56. S. L. Smith, *Am. Lab. (Fairfield, Conn.)*, *17* (11): 82 (1985).

57. S. L. Smith, *Appl. Spectrosc.*, *40*: 278 (1986).

58. T. Hirschfeld, *Anal. Chem.*, *52*: 297A (1980).

59. C. L. Wilkins, G. N. Giss, G. M. Brissey, and S. Steiner, *Anal. Chem.*, *53*. 113 (1981).

60. R. W. Crawford, T. Hirschfeld, R. H. Sanborn, and C. M. Wong, *Anal. Chem.*, *54*: 817 (1982).

61. C. L. Wilkins, G. N. Giss, R. L. White, G. M. Brissey, and E. C. Onyiriuka, *Anal. Chem.*, *54*: 2260 (1982).

62. C. L. Wilkins, *Science*, *222*: 291 (1983).

63. D. A. Laude, Jr., G. M. Brissey, C. F. Ijames, R. S. Brown, and C. L. Wilkins, *Anal. Chem.*, *56*: 1163 (1984).

64. D. A. Laude, Jr., C. Johlman, and C. L. Wilkins, *Opt. Eng.*, *24*: 1011 (1985).

65. J. R. Cooper, I. C. Bowater, and C. L. Wilkins, *Anal. Chem.*, *58*: 2791 (1986).

66. R. J. Leibrand, *Am. Lab. (Fairfield, Conn.)*, *20* (12): 40 (1988).

67. S. S. Williams, R. B. Lam, D. T. Sparks, T. L. Isenhour, and J. R. Hass, *Anal. Chim. Acta, 138*: 1 (1982).

68. S. S. Williams, R. B. Lam, and T. L. Isenhour, *Anal. Chem.*, *55*: 1117 (1983).

69. *Chem. Engr. News*, *62* (4): 14 (1984).

70. D. A. Laude, Jr., C. L. Johlman, J. R. Cooper, and C. L. Wilkins, *Anal. Chem.*, *57*: 1044 (1985).

3
Liquid Chromatography/
Fourier Transform Infrared Spectroscopy

I. INTRODUCTION

Nonvolatile and thermally labile mixture components that cannot
be isolated by gas chromatography can often be separated by
using high-pressure liquid chromatography (HPLC) or supercritical fluid chromatography (SFC). In contrast to GC, infrared-transparent atomic and homonuclear diatomic mobile phases
are inappropriate for HPLC and SFC separations. Unfortunately,
infrared absorptions for common HPLC and SFC mobile phases
often overlap eluent absorbance bands, making detection and
identification difficult. It is mobile phase opacity that is primarily responsible for the lack of general-purpose HPLC/FT-IR
and SFC/FT-IR interfaces possessing sensitivity comparable to
GC/FT-IR.

Liquid chromatographic separations can be categorized by
eluent/stationary phase interactions. In adsorption chromatography, mixture components are separated by competition between
eluents and mobile phase for stationary phase active sites.
Silica is commonly employed as an adsorption chromatography

stationary phase. Silica contains a large number of polar active sites on which eluents can adsorb during elution. The nature of eluent/stationary phase interactions can be altered by modifying the silica surface. Bonded phase chromatographic packings are produced by chemically attaching organic functionalities to polar silica sites. Both polar and nonpolar bonded phase packings have been synthesized in this manner. The most popular nonpolar bonded phase packings contain either C_8 or C_{18} organic chains attached to a silica surface. Separation mechanisms involved in adsorption chromatography differ significantly from bonded phase HPLC separations. Two classes of HPLC separation (normal and reverse phase) have been established to distinguish these mechanisms. Normal phase separations (adsorption) employ polar stationary phases and nonpolar mobile phases. Reverse phase separations employ nonpolar stationary phases and polar mobile phases. About 60% of all current HPLC separations employ reverse phase chromatography [1]. Unfortunately, common reverse phase solvents (e.g., H_2O, CH_3OH, CH_3CN) are highly infrared-absorbing and pose particularly difficult challenges for FT-IR detection.

In size exclusion (gel permeation) chromatography, mixture components are separated based on molecular shape. The stationary phase contains pores of varying sizes and shapes. Small molecules can become trapped in the pores and retained on the column longer than larger molecules. Ionic species can be separated by using ion-exchange and ion-pair chromatography. In ion-exchange chromatography, ion separation is based on differing affinity for charged sites exposed on the surface of the stationary phase. In ion-pair chromatography, ion polarity is reduced by adding counter-ions to the mobile phase. Separation of neutral ion-pair complexes is based on differing molecular functionalities with minimal effects from ionic charge.

By using small-particle-size supports, chromatographic columns can be made smaller and separating efficiencies can be increased. Microbore HPLC offers higher chromatographic resolution than analytical-scale HPLC. In addition, solvent expenditures are reduced for microbore HPLC because column volumes are much smaller and flow rates are significantly reduced. HPLC/FT-IR interfaces have been developed for all HPLC categories except ion exchange.

Some of the advantages of gas chromatography can be retained in nonvolatile mixture-component separation by employing

supercritical fluid chromatography. When supercritical conditions exist, SFC mobile phase is converted from a gas into a
fluid. Supercritical fluid solvation characteristics are intermediate between gaseous and liquid states. Thus, eluent solvation similar to HPLC can be achieved with high-resolution GC
columns at temperatures that are low enough to prevent thermal
degradation.

II. BRIEF HISTORY

The first interface between liquid chromatography and FT-IR
was described by Kizer et al. in 1975 [2]. A UV detector was
used to trigger FT-IR data acquisition when chromatographic
eluents entered a 0.030 mm path length AgCl flow cell. Detection limits for this interface were in the high microgram range.
The low sensitivity of this HPLC /FT-IR combination was primarily due to the immaturity of FT-IR technology at the time.
Subsequently, increased interferometer scan rate and reproducibility led to higher FT-IR photomeric accuracy (improved
spectral subtractions), and more sensitive infrared detectors
(MCT) were developed that had short response times, leading
to better chromatogram digitization. With these improvements,
Vidrine and Mattson refined flow cell HPLC /FT-IR interfaces
and studied the dependence of detection limit on mobile phase
opacity [3]. They attained submicrogram detection limits with
CCl4 mobile phase.
        The first mobile phase elimination HPLC /FT-IR interface
consisted of spraying chromatographic effluent into heated
light pipes [4]. This approach was unsuccessful because
eluent could not be deposited evenly over the internal surface
of the light pipe. Other mobile phase removal attempts included a moving-band interface similar to LC /MS designs [5]
and a diffuse reflectance FT-IR (DRIFTS) carousel [6]. The
diffuse reflectance carousel yielded the best results and was
shown to be compatible with gradient elution chromatography.
Detection limits of 100 ng were reported for the diffuse reflectance carousel interface combined with analytical-scale HPLC.
        FT-IR detection was first applied to the analysis of micro-
bore HPLC eluent by Teramae and Tanaka [7]. Microbore
HPLC provides greater chromatographic resolution than conventional analytical-scale HPLC. In addition, exotic mobile phases
can be employed for microbore HPLC /FT-IR because only small
amounts of solvent are consumed during separation [8–14].

The first SFC/FT-IR interface was reported by Shafer and
Griffiths in 1983 [15]. SFC/FT-IR detection limits below 50 ng
have been obtained by using a mobile phase elimination interface [16].

III. FLOW CELL INTERFACE

The flow cell HPLC/FT-IR interface is a logical application of
the liquid transmission cell to flowing solvents. Flow cells consist of two infrared-transparent windows separated by a spacer
(Figure 3-1). The thickness of the spacer (1 mm teflon pad
in Figure 3-1) determines the path length and volume of the
flow cell. Potassium bromide windows can be used with nonaqueous mobile phases. Silver chloride and zinc selenide
windows have been employed with aqueous solvents.

A. Path Length Considerations

Theoretical optimum dimensions for HPLC/FT-IR flow cells can
be estimated in the same manner described for GC/FT-IR light
pipes (Chapter 2, p. 47). A typical analytical-scale HPLC
column with a length of 250 mm and 20,000 theoretical plates
yields elution profiles with $V_{P_{\frac{1}{2}}}$ of about 76 µl [18]. For a 3
mm beam diameter (same size as was used for GC/FT-IR light
pipe estimates), the optimum cell path length would be 11 mm.
Unfortunately, infrared absorbance of commonly employed
mobile phases precludes the use of flow cells with path lengths
this long. Flow cell path lengths of 0.1—0.2 mm are commonly
employed with moderately absorbing mobile phases (e.g.,
n-hexane). For highly absorbing mobile phases such as water,
flow cell path length must be reduced to about 10 µm. From
Beer's law, flow cell detection limits for reverse phase HPLC/
FT-IR can be predicted to be 1000 times higher than if the
mobile phase were infrared-transparent.

Mixture-component infrared spectra are obtained from flow
cell measurements by subtracting mobile phase contributions
from acquired spectra. Reported detection limits for flow cell
HPLC/FT-IR applications can sometimes be misleading. Often,
only a single infrared absorption can be discerned from difference spectra obtained at the detection limit. Combellas et
al. investigated variations in spectral quality for varying quantities of anisole in CCl_4 mobile phase [19]. The number of
anisole infrared bands that could be discerned from spectra

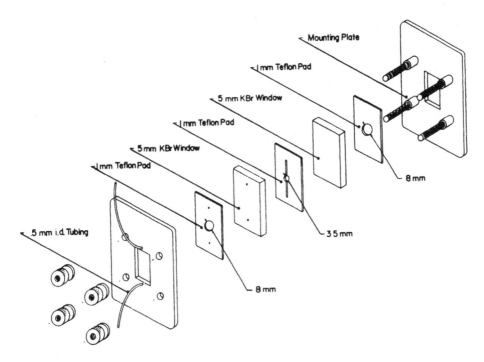

Figure 3-1   Exploded view of a transmission flow cell HPLC /
FT-IR interface.   (Reprinted with permission from Ref. 17.
Copyright © 1979 Academic Press, New York.)

was dependent on the amount of anisole injected (Table 3-1).
The detection limit was specified at 500 ng.   However, 300 µg
was required in order to obtain a comprehensive spectrum.

The difficulty in removing mobile phase absorbance features
from HPLC /FT-IR spectra by absorbance subtraction can be
explained by considering relative concentration error (σ_C/C)
as a function of solvent transmittance.   This relation can be
written as [20]

$$\frac{\sigma_C}{C} = \frac{\sigma_T}{T \log T \ln 10} \tag{3.1}$$

where C is concentration, σ_C the error in calculated concentration, T the transmittance of the solution, and σ_T the error in

Table 3-1   Number of Infrared Bands
Observed for Anisole in CCl_4

Amount injected (μg)	Number of IR bands
300	15
150	10
75	6
30	4
10	2
0.5	1

Source: Reprinted with permission from Ref.
19. Copyright © 1983 Elsevier Science Publishers, New York.

transmittance measurements. This function is shown in
Figure 3-2 for a transmittance error (σ_T) of 0.1 %T. The
optimum %T for spectral measurements occurs at 36.8 %T
(0.434 A.U.), but relative error is acceptable over a range of
10—80 %T (0.1—1 A.U.). In order to facilitate removal of
mobile phase absorbance contributions without significantly increasing spectral noise in difference spectra, mobile phase absorbance should be confined to a range from 0.1—1 A.U. over
infrared regions for which information is desired. Common
HPLC mobile phases exhibit characteristic infrared spectra;
therefore, mobile phase absorptivity varies with wavelength.
The optimum HPLC/FT-IR path length at a given wavelength is
inversely proportional to the absorptivity of the mobile phase
at that wavelength. Figure 3-3 contains plots of optimum path
length (path length at which absorbance is equal to 0.434 A.U.)
as a function of wavelength for some common HPLC mobile
phases. Carbon tetrachloride [Figure 3-3(a)] is infrared-
transparent over a wide range of infrared wavelengths and is
often employed as a solvent for size-exclusion HPLC/FT-IR
separations. A common mobile phase for normal phase chromatography is *n*-hexane [Figure 3-3(b)]. Path lengths of about
0.2 mm yield information over most of the infrared range.
Reverse phase solvents typically contain water [Figure 3-3(c)].

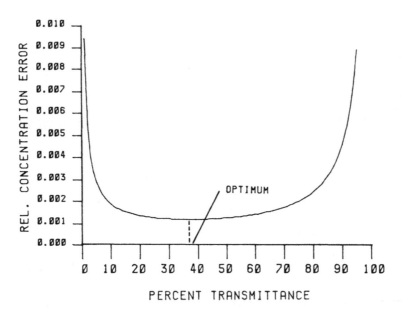

Figure 3-2   Variation of relative concentration error as a function of transmittance.

Path lengths should not exceed 20 μm for HPLC /FT-IR flow cell measurements when aqueous mobile phases are employed.

Detection limits approaching those for GC /FT-IR light pipe interfaces have been reported for flow cell HPLC /FT-IR when infrared-transparent mobile phases are employed. For example, in one of the first HPLC /FT-IR reports, Vidrine and Mattson used CCl_4 mobile phase with a 11.2 mm path length flow cell to detect 525 ng of paraffin oil eluting from a HPLC column [3] (Figure 3-4).

Most early HPLC /FT-IR reports involved applications of size-exclusion chromatography for which solvents such as CCl_4 and $CHCl_3$ could be employed [3,21—23]. Later, normal phase and reverse phase flow cell interfaces were developed [24—26]. Size-exclusion mobile phases serve only to dissolve the mixture. Therefore, infrared-transparent solvents can be employed without degrading chromatographic efficiency.

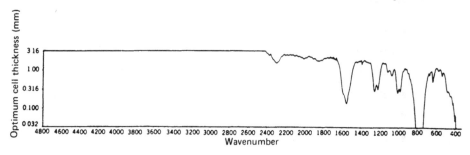

(a)

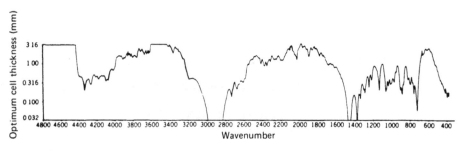

(b)

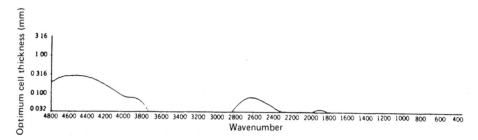

(c)

Figure 3-3   Optimum flow cell path lengths as a function of wavelength for (a) carbon tetrachloride, (b) n-hexane, and (c) water mobile phases.   (Reprinted with permission from Ref. 17.   Copyright © 1979 Academic Press, New York.)

Mobile phases commonly employed for reverse phase HPLC exhibit intense infrared absorptions and are incompatible with FT-IR detection. Johnson et al. recently described a method in which reverse phase solvents could be used for chromatographic separation and then replaced with infrared-transparent solvents prior to FT-IR detection [26]. Their segmented-flow continuous-extraction interface is shown in Figure 3-5. Column effluent (pump A) is mixed with extraction solvent (pump B) in the extraction coil (E). This mixture is passed to a hydrophobic membrane (F) (Figure 3-6) that separates aqueous (CH_3OH/H_2O) and nonaqueous solvents. The nonaqueous solvent ($CHCl_3$ or CCl_4) is directed to the FT-IR flow cell (G) for analysis. Carbon tetrachloride was found to be a better extraction solvent than chloroform. Carbon tetrachloride dissolves a wide variety of solutes, absorbs less water, and has greater infrared transparency than chloroform. Chromatograms in Figures 3-7 and 3-8 illustrate the performance of the continuous-extraction interface with chloroform and carbon tetrachloride mobile phases for a mixture of cyclohexanone, acetophenone, and benzophenone. The SNR for the chromatogram in Figure 3-8 is significantly greater than that of the chromatogram in Figure 3-7. In fact, benzophenone was not detected when chloroform mobile phase was employed (Figure 3-7). This illustrates the need for eluent distribution coefficients that favor the extraction solvent for this type of interface.

B. Cylindrical Transmission Flow Cell

Transmission flow cells considered in the previous section are constructed by compressing two infrared-transparent windows about a spacer used to select path length. Johnson and Taylor recently introduced a new flow cell design that permits simultaneous transmittance measurements for multiple path lengths [27]. The cylindrical flow cell can be fabricated by simply boring a hole through infrared transmitting material. One advantage of the cylindrical flow cell is that the HPLC column can be connected directly to the infrared sampling region with essentially no dead volume (Figure 3-9). For this reason, this type of flow cell has been called "zero dead volume" (ZDV). Because ZDV flow cell geometry is cylindrical, path length varies continuously from zero to the diameter of the cylinder (Figure 3-10). The following expression can be

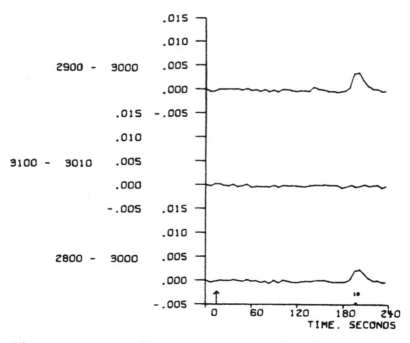

(a)

Figure 3-4   Adsorption chromatography separation of paraffin
oil.   (a) Integrated absorbance chromatogram.   (b) Paraffin
oil infrared spectrum.   (Reprinted with permission from Ref. 3.
Copyright © 1978 Society for Applied Spectroscopy, Frederick,
Maryland.)

derived to describe the effective absorbance for eluents contained in a cylindrical flow cell:

$$A = - \log \frac{\int_{-r}^{r} 10^{-ab(x)C}}{2r} \, dx \qquad (3.2)$$

In the above equation, A is the measured absorbance of the
flow cell, a the absorptivity of the species in the cell, C its
concentration, b the cell path length that is a function of the
position of the entrance beam (x), and r the radius of the
cylindrical flow cell.

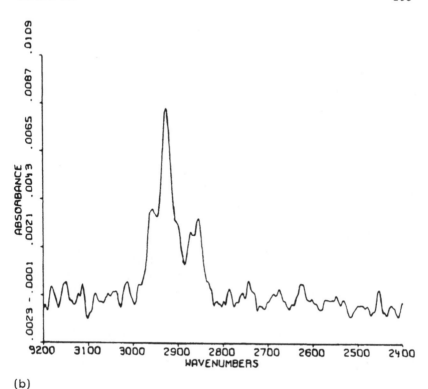

(b)

Figure 3-4   (Continued)

The range of concentrations that can be measured with adequate SNR (dynamic range) with cylindrical flow cells is greater than fixed path length cells. Dynamic range enhancement for multiple path length transmission cells was first suggested by Hirschfeld [28] and later demonstrated by Dasgupta [29]. To examine the effect of multiple path length on spectral measurements, consider the simple case of a transmission cell comprised of three different path length regions. If we assume that one-third of the incident radiation enters each of the three regions and that the cell contains a typical organic absorber (e.g., $a = 10 \ M^{-1} \ cm^{-1}$), contributions to the total amount of transmitted radiation from the three regions with path lengths of b, b/10, and 10b (b = 0.1 cm) can be

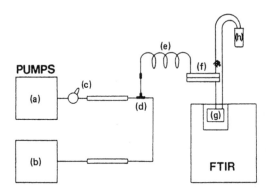

Figure 3-5   Diagram of segmented flow interface for reverse phase HPLC/FT-IR.   (a) HPLC pump; (b) extraction solvent pump; (c) HPLC injector; (d) segmenting tee; (e) extraction coil; (f) membrane phase separator; (g) FT-IR flow cell interface; and (h) waste.   (Reprinted with permission from Ref. 26. Copyright © 1985 American Chemical Society, Washington, D.C.)

represented, as shown in Figure 3-11.   At high concentration, most of the transmitted radiation comes from the short path length region and essentially no radiation emerges from the other regions.   At low concentration, transmitted radiation contributions from each path length region are approximately equal (i.e., nearly all radiation is transmitted by all three regions). Calculated absorbance for the three region cell as a function of

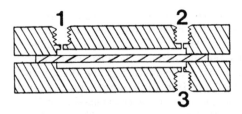

Figure 3-6   Cross-sectional view of membrane separator shown in Figure 3-5.   (1) Segmented stream inlet; (2) aqueous waste outlet; and (3) organic phase outlet.   (Reprinted with permission from Ref. 26.   Copyright © 1985 American Chemical Society, Washington, D.C.)

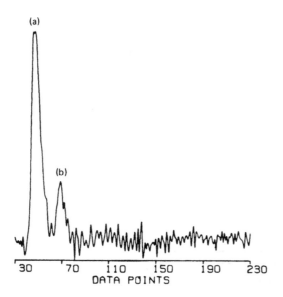

Figure 3-7   Gram—Schmidt-reconstructed chromatogram for a
reverse phase HPLC/FT-IR separation of (a) cyclohexanone;
(b) acetophenone; and (c) benzophenone (not detected) using
CHCl₃ mobile phase.   (Reprinted with permission from Ref. 26.
Copyright © 1985 American Chemical Society, Washington, D.C.)

eluent concentration is illustrated in Figure 3-12.   The straight
line represents the response expected if the path length of the
flow cell were simply b.   Figure 3-12 shows that the dynamic
range for the multiple path length cell is greater than for the
single path length cell.   However, the slope of the absorbance
vs concentration curve (i.e., sensitivity) for the multiple path
length cell over a specified concentration range is less than
that for the single path length cell.   Also, the absorbance vs
concentration curve shown in Figure 3-12 does not obey Beer's
law [see Eq. (3.2)].   Therefore, the cylindrical flow cell provides dynamic range enhancement at the expense of sensitivity
and Beer's law linearity.   These factors should be considered
when attempting to make quantitative measurements with a
cylindrical flow cell.
    Effects of extended dynamic range derived from cylindrical
cell measurements on spectral subtraction results are illustrated

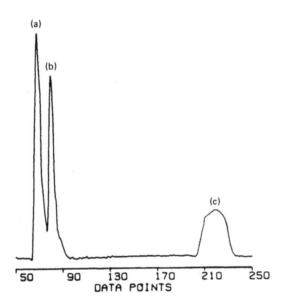

Figure 3-8 Gram—Schmidt-reconstructed chromatogram for a reverse phase HPLC/FT-IR separation of (a) cyclohexanone; (b) acetophenone; and (c) benzophenone using CCl_4 mobile phase. (Reprinted with permission from Ref. 26. Copyright © 1985 American Chemical Society, Washington, D.C.)

in Figure 3-13 for 400 ppm anisole in CCl_4. Figure 3-13(a) is a difference spectrum obtained by subtracting the spectrum of CCl_4 from an anisole mixture spectrum measured by using a 0.75 mm diameter cylindrical transmission cell. Figure 3-13(b) is a difference spectrum for the same mixture obtained from conventional flow cell measurements by using a 0.5 mm fixed path length cell. Figure 3-13(c) is a reference spectrum of CCl_4 solvent. The difference spectrum for the fixed path length cell contains greater noise in spectral regions where the mobile phase was highly absorbing (>1 A.U.). This noise leads to the spectral distortions shown in Figure 3-13(b) that occur near CCl_4 absorbance maxima. The extended dynamic range provided by the cylindrical cell minimizes this noise in difference spectra [Figure 3-13(a)]. The SNR of the cylindrical cell difference spectrum is greater than that of the fixed path

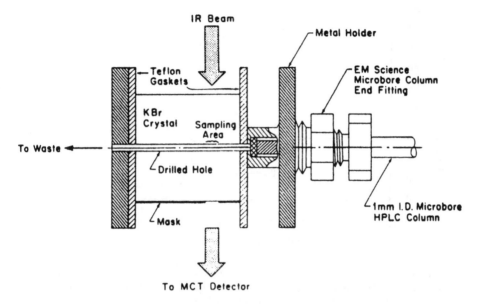

Figure 3-9   Cylindrical transmission flow cell coupled to a microbore HPLC column. (Reprinted with permission from Ref. 27. Copyright © 1984 American Chemical Society, Washington, D.C.)

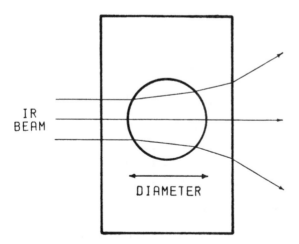

Figure 3-10   Cross-sectional view of a cylindrical flow cell illustrating multiple path lengths.

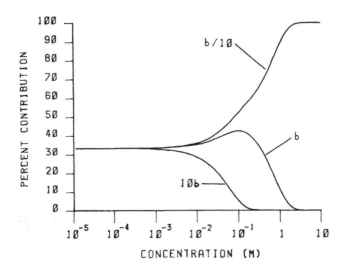

Figure 3-11 Theoretical variation of transmitted radiation contributions from each region of a three-path-length transmission
cell as a function of eluent concentration.

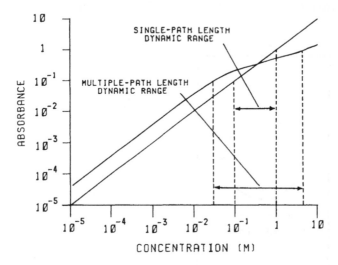

Figure 3-12 Theoretical absorbance for a three-region
multiple path length flow cell and a single path length cell.

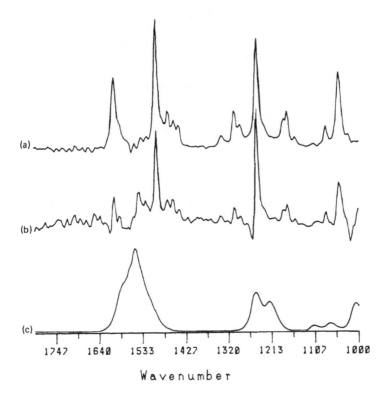

Figure 3-13 (a) Difference spectrum for 400 ppm anisole in CCl$_4$ obtained by using a cylindrical transmission cell. (b) Difference spectrum for 400 ppm anisole in CCl$_4$ obtained by using a conventional transmission flow cell. (c) Reference spectrum of CCl$_4$.

length flow cell. This is consistent with the theoretical model of cylindrical cell absorbance (Figure 3-12) that predicts higher absorbance than conventional flow cell measurements at low concentration.

C. Cylindrical Internal Reflectance Cell

With some difficulty, fixed path length transmission flow cells can be constructed for aqueous solution analysis. These cells

are easily clogged and internal pressure fluctuations can
change path length appreciably. Attenuated total reflectance
(ATR) provides an alternative method for aqueous solution
analysis that does not suffer from these problems. For ATR
analysis, sample material is placed in contact with a high-
refractive-index (n_{ATR}) infrared-transparent material. ZnSe
is often employed for aqueous solution analysis. When incident
radiation strikes the material at an angle greater than the
critical angle (θ_C), total internal reflection occurs.

$$\theta_C = \sin^{-1} (n_{ATR}/n_{SAMPLE}) \qquad (3.3)$$

At each internal reflection, the electric field of the infrared
radiation extends slightly into the sample material. This penetration is known as an evanescent wave. The evanescent wave
interacts with sample material and is either absorbed or reflected back into the ATR material. The depth of penetration
(d_p) of the evanescent wave into the sample material (path
length) depends on the refractive indices of the materials at
the ATR boundary (n_{ATR}, n_{HPLC}), the angle of incidence (θ),
and the wavelength of infrared radiation (λ) [30].

$$d_p = \frac{\lambda}{2\pi [\sin^2 (\theta) - (n_{HPLC}/n_{ATR})^2]^{1/2}} \qquad (3.4)$$

An ATR flow cell constructed from a cylindrical ZnSe rod is
particularly well matched to the circular beam profile of FT-IR
spectrometers [31] (Figure 3-14).

Sabo et al. reported the first application of an ATR flow
cell to HPLC/FT-IR analysis [32]. They used the ATR flow
cell for both normal phase (methylene chloride mobile phase)
and reverse phase (60:40 v/v methanol:water mobile phase)
HPLC/FT-IR. Ten reflections were obtained in the ATR cell
that yielded a path length of 4–22 µm (path length varies
with λ) over the mid-infrared range. Due to the extremely
short path length, detection limits for this interface were poor
($\sim$1 mg/component). Although reverse phase HPLC/FT-IR
interface designs are simplified by using an ATR flow cell, the
short path length obtained (10–20 µm) results in detection
limits that are more than 1000 times higher than the theoretical
optimum. For normal phase separations, transmission flow
cells perform better than ATR cells because longer path
lengths can be employed.

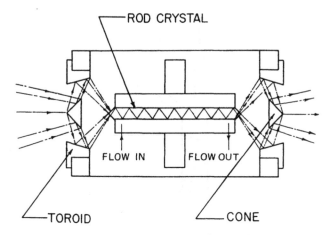

Figure 3-14   Schematic of a cylindrical rod ATR HPLC /FT-IR flow cell.

IV. MOBILE PHASE ELIMINATION

Kuehl and Griffiths developed the first practical mobile phase elimination HPLC /FT-IR interface in 1979 [6,33]. This inter-face consisted of a solvent concentrator (Figure 3-15) that evaporated about 90% of the mobile phase and a series of dif-fuse reflectance sample cups arranged in a carousel (Figure 3-16). Each sample cup in the carousel was filled with infra-red transparent substrate (KCl). The signal from a UV detec-tor inserted between the HPLC and FT-IR was used to initiate deposition of separated mixture components on diffuse reflectance substrate. This was accomplished by closing the solenoid valve to the aspirator shown in Figure 3-15. After deposition, the diffuse reflectance sample cup was rotated to a second carousel position where the remainder of the HPLC solvent was evaporated. Once all mobile phase had been eliminated, isolated mixture component infrared spectra could be measured by DRIFTS (Figure 3-16). The diffuse reflectance carousel interface com-bined seemingly incompatible features of confined eluent deposi-tion (4.5 mm diameter sample cup) with a large surface area for solvent evaporation (50 cm^2 surface area for powdered KCl substrate). The infrared beam for DRIFTS analysis typically penetrates to a depth of 1–2 mm for powdered KCl substrate. This was adequate because at least 75% of all deposited eluent

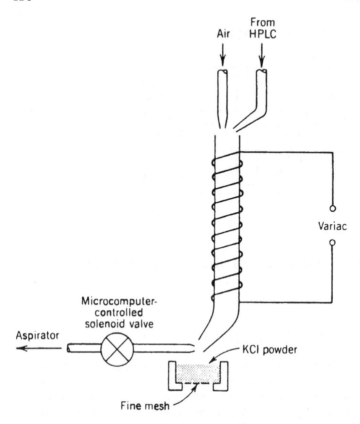

Figure 3-15 Solvent concentrator for DRIFTS mobile phase elimination HPLC/FT-IR interface. (Reproduced from Ref. 6 by permission of Preston Publications, a division of Preston Industries, Inc., Niles, Illinois.)

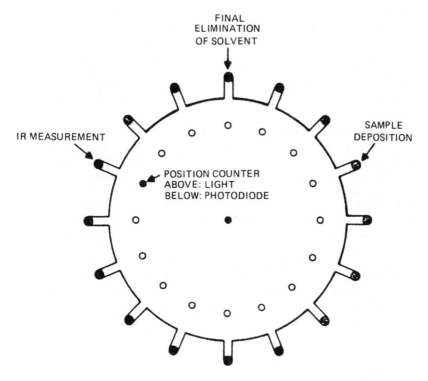

Figure 3-16   Diffuse reflectance carousel mobile phase elimination HPLC/FT-IR interface. (Reproduced from Ref. 6 by permission of Preston Publications, a division of Preston Industries, Inc., Niles, Illinois.)

was found to reside on the surface of the KCl. DRIFTS analysis has been shown to be more sensitive than corresponding KBr pellet transmission measurements [34]. High enrichment factors attained for the DRIFTS interface due to the removal of HPLC mobile phase result in low detection limits (∿100 ng) for this interface.

The first diffuse reflectance carousel HPLC/FT-IR interface suffered from several drawbacks. A UV detector placed between the HPLC and FT-IR was required to detect component elutions and trigger eluent deposition. Thus, continuous infrared monitoring of HPLC effluent was not possible. Chromatogram interrogation was limited to 32 chromatographic eluents (i.e., the number of sample cups). High detection limits were obtained for moderately volatile mixture components because some eluent was lost during mobile phase elimination. In addition, the ultimate sensitivity of the diffuse reflectance interface was dictated by impurities contained in the KCl substrate and mobile phase solvent (chemical noise) and not by instrumental noise.

A variation of the diffuse reflectance carousel was developed for the analysis of reverse phase HPLC eluents [35]. The design was similar to an HPLC/MS interface in which aqueous eluents are extracted into CH_2Cl_2 prior to deposition on a moving-belt LC/MS interface [36]. HPLC effluent and CH_2Cl_2 extraction solvent were continuously mixed in an extraction coil (Figure 3-17). The lighter aqueous phase was removed by aspiration and the CH_2Cl_2 extraction solvent was passed to an eluent concentrator. Some modifications of the original diffuse reflectance carousel were made to facilitate continuous operation and eliminate the need for an auxiliary HPLC detector. The carousel used for this interface contained 160 sample positions and effluent was deposited at each position for 15 sec (40 min total analysis time available). Potassium chloride pellets were

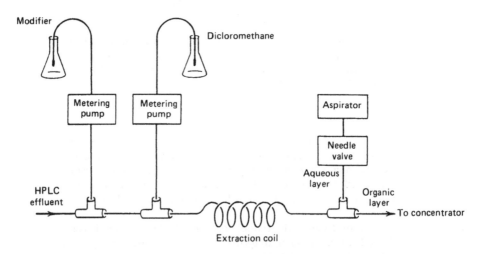

Figure 3-17   Diagram of continuous extraction reverse phase HPLC/FT-IR interface. (Reprinted with permission from Ref. 35. Copyright © 1984 American Chemical Society, Washington, D.C.)

used for DRIFTS substrate instead of KCl powder because pellets could be fabricated more easily and reproducibly. Detection limits below 1 µg were obtained with this interface for reverse phase HPLC/FT-IR.

V. MICROBORE HPLC/FT-IR

Microbore HPLC can be employed to obtain greater chromatographic resolution than conventional analytical-scale HPLC. Microbore HPLC has some properties that are especially attractive for HPLC/FT-IR. The amount of mobile phase required for separation is much less than for conventional analytical-scale HPLC. In addition, optimum flow rates for microbore HPLC are lower than those for analytical-scale HPLC. These two features simplify both flow cell and mobile phase elimination HPLC/FT-IR interfaces. Low mobile phase consumption for microbore HPLC permits the cost-effective use of exotic mobile phases that have high infrared transmittance over useful spectral ranges. Also, microbore HPLC/FT-IR mobile phase elimination interfaces can be designed in such a way as to continuously remove solvent by evaporation, reaction, or extraction.

A. Flow Cell Interface

Flow cell microbore HPLC/FT-IR interfaces have been developed that employ solvents such as $CDCl_3$ [8,11,12,14], Freon-113 [9], D_2O [10,14], and CD_3CN [14]. Deuterated solvents have infrared absorptivities that are similar to their hydrogen analogs. However, regions of high absorbance are shifted in deuterated solvent spectra due to the isotope effect of deuterium on molecular vibrations. Thus, eluent C—H stretching vibrations are not overlapped by mobile phase absorptions when $CDCl_3$ and D_2O are employed. This makes flow cell HPLC/FT-IR detection of reverse phase eluents feasible. Detection limits in the low µg range have been reported for flow cell reverse phase HPLC/FT-IR using $CD_3CN:D_2O$ (90:10 v/v) mobile phase [14].

B. Mobile Phase Elimination Interface

Both normal and reverse phase microbore HPLC/FT-IR solvent elimination interfaces have been constructed. Detection limits for these interfaces are dictated by the concentration of impurities in the mobile phase, and the useful dynamic range is limited by the sample capacity of microbore columns. HPLC column capacity decreases in proportion to cross-sectional area. Thus, the capacity of a 1 mm diameter microbore column is approximately 20 times less than a 4.6 mm diameter (analytical-scale) column.

1. Normal Phase

The diffuse reflectance carousel HPLC/FT-IR interface described previously was modified for use with microbore HPLC/FT-IR by reducing the sample cup diameter from 4.5 to 2.5 mm and expanding the number of sample cups to 180 [37]. Because microbore HPLC flow rates are much lower than those used for analytical-scale HPLC, the mobile phase concentrator was not required. Detection limits in the low nanogram range were obtained with this interface.

A *buffer memory* microbore HPLC/FT-IR interface was designed by Jinno and co-workers [38–41]. Their interface employed transmission spectroscopy and a continuously moving KBr window (Figure 3-18). HPLC effluent was continuously sprayed onto the KBr window where the mobile phase evaporated, leaving behind mixture components. Isolated components were located at positions on the infrared-transparent window that were representative of their respective retention times. Eluent was deposited as a continuous narrow band approximately 1.5 mm in width at a deposition rate of 1.25 mm/min. After separation, the window was passed through the infrared beam of an FT-IR in order to record a HPLC/FT-IR chromatogram. By using a beam condenser, spectra were obtained at 0.5 mm intervals on the window that corresponded to a temporal resolution of 24 sec.

An HPLC/FT-IR interface based on a similar concept was described by Gagel and Biemann [42]. Instead of depositing HPLC effluent on an infrared-transparent substrate, a highly reflective surface was employed (Figure 3-19). Infrared spectra were obtained by reflecting an infrared beam from the deposition surface (Figure 3-20). Detection limits for buffer-memory-type transmission and reflection HPLC/FT-IR interfaces are about the same ($\sim$100 ng). Evidently, detection limits are dictated by the concentration of impurities in the mobile phase and not by the method of obtaining infrared spectra.

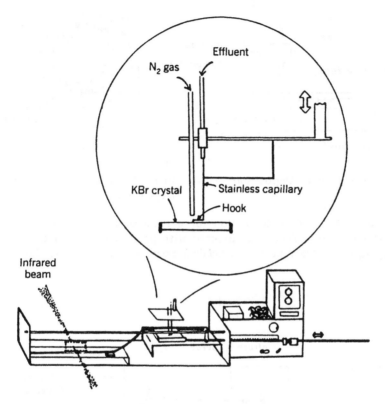

Figure 3-18 Buffer memory microbore HPLC /FT-IR interface. (Reprinted with permission from Ref. 40. Copyright © 1982 Society for Applied Spectroscopy, Frederick, Maryland.)

2. *Reverse Phase*

Normal phase microbore HPLC /FT-IR mobile phase elimination interfaces work well when solvent is easily evaporated. This is not the case for reverse phase solvents that typically contain water. Kalasinsky et al. developed an HPLC /FT-IR interface in which water was removed via chemical reaction with 2,2-di-methoxypropane (DMP) [43]

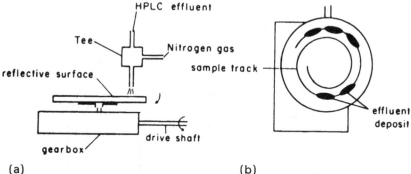

(a)
Side View During Deposition

(b)
Front View with Components
Deposited

Figure 3-19   Eluent deposition for a reflection-absorption
HPLC/FT-IR interface. (Reprinted with permission from Ref. 42.
Copyright © 1986 American Chemical Society, Washington, D.C.)

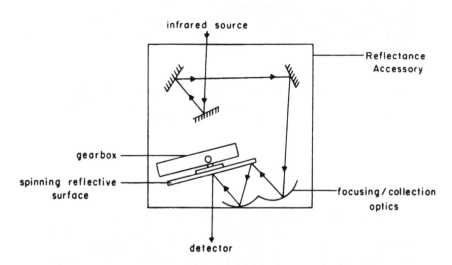

Figure 3-20   FT-IR optical interface for reflection-absorption
HPLC/FT-IR interface. (Reprinted with permission from Ref. 42.
Copyright © 1986 American Chemical Society, Washington, D.C.)

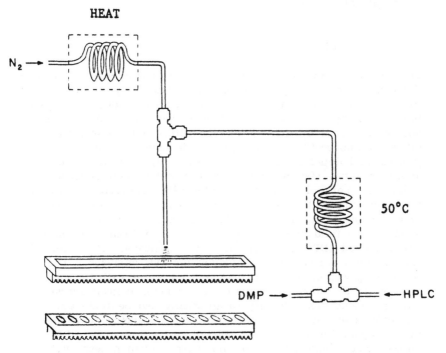

HEAT

$N_2 \rightarrow$

50°C

DMP → ← HPLC

Figure 3-21 Diagram of a mobile phase elimination HPLC/FT-IR interface for reverse phase separations employing continuous reaction of water with DMP. (Reprinted with permission from Ref. 43. Copyright © 1985 American Chemical Society, Washington, D.C.)

$$DMP + H_2O = 2CH_3OH + (CH_3)_2CO$$

Reaction products (methanol and acetone) are more easily evaporated than water. Mixture components can be deposited on diffuse reflectance substrate in a manner similar to that previously described for the diffuse reflectance carousel interface (Figure 3-21). Complete water removal has been achieved for mobile phases containing up to 80% H_2O.

VI.   SFC/FT-IR

Supercritical fluid chromatography (SFC) is applicable to separations in which mobile phase properties between typical GC and HPLC solvents are desired. It is particularly useful for separations involving nonvolatile or thermally labile materials that cannot be separated by GC. Fluid viscosities and solute diffusivities for supercritical mobile phases lie between those of gaseous and liquid states. As a result, higher mobile phase linear velocities (faster analysis times) and greater separation efficiencies per unit time are possible with supercritical chromatography compared to HPLC. Both GC and HPLC equipment (e.g., columns and detectors) have been used for SFC [44,45]. Supercritical fluid chromatography can be performed by using either packed (HPLC) or capillary (GC) columns. HPLC pumps and injectors are commonly employed for SFC analyses. The solvating power of SFC mobile phases increases with increasing density (i.e., column pressure). By varying column pressure during a separation, density programming can be achieved. The effects of SFC density programming are similar to temperature programming for GC and solvent programming for HPLC. Common SFC mobile phases are nonpolar. When necessary, small quantities of polar modifiers such as methanol can be added to SFC mobile phases to increase solubility of polar species in the mobile phase.

Hirschfeld suggested coupling supercritical fluid chromatography with infrared detection in 1980 [46] and the first SFC/FT-IR interface was described by Shafer et al. in 1983 [15]. Carbon dioxide is most commonly employed for SFC/FT-IR mobile phase. However, applications involving n-hexane [47] and xenon [48] mobile phases have also been reported. SFC/FT-IR interfaces have been devised using flow cells, mobile phase elimination, and matrix isolation methodologies.

A. Flow Cell Interface

Infrared spectra for gaseous, supercritical, and liquid CO_2 are
shown in Figure 3-22. Absorbance bands at 1286 cm^{-1} and
1388 cm^{-1} in the spectrum of supercritical CO_2 [Figure 3-22(c)]
arise from Fermi resonance between a Raman-active symmetric
stretch at 1300 cm^{-1} and the second harmonic of the doubly
degenerate bending vibration at 667 cm^{-1} [49]. With the exception of O—H stretching vibrations, infrared absorbance
bands for most organic eluents are not obscured by CO_2 absorptions. In this respect, CO_2 is nearly an ideal SFC/FT-IR
mobile phase. The infrared transparency of CO_2 permits the
use of flow cells with path lengths 10—100 times greater than
those employed for HPLC/FT-IR. For instance, the first interface between SFC and FT-IR employed a 10 mm path length
flow cell with CaF_2 windows [15]. Subsequent flow cell designs
utilized path lengths between 5 and 10 mm [50—54]. A critical
component of SFC/FT-IR flow cells are the infrared transparent
windows. These windows must withstand operating pressures
of 2000—4000 psi. The minimum thickness of infrared transparent windows used for flow cell construction can be derived
from the following expression [52]:

$$\text{Thickness}_{(MIN)} = [(1.1 \ P \ r^2 \ S)/M_R]^{1/2} \qquad (3.5)$$

where P is the maximum operating pressure (psi), r the radius
of unsupported aperture (mm), S a safety factor, and M_R the
modulus of rupture for the window material (psi). Values of
M_R for common infrared-transparent windows are compiled in
Table 3-2. SFC/FT-IR flow cells typically incorporate infrared-
transparent windows with 2 mm thicknesses. These flow cells
are subject to the pressure and geometric restrictions implied
by Eq. (3.5).

SFC/FT-IR flow cells can be made from high-pressure UV
flow cells by simply replacing the quartz windows with appropriate infrared-transparent windows. Alternatively, light pipes
similar to GC/FT-IR designs have been employed for SFC/FT-IR
[53]. An example of a light pipe SFC/FT-IR flow cell is illustrated in Figure 3-23. The gold internal surface of the flow
cell provides higher infrared radiation throughput for long
path length cells compared to modified UV flow cells.

Supercritical fluid density and solvent strength depend on
the pressure of the fluid. By changing the supercritical fluid

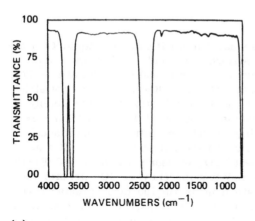

(a)

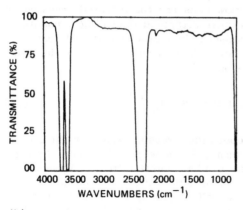

(b)

Figure 3-22   Infrared spectra of CO_2 in (a) gaseous state;
(b) liquid state; and (c) supercritical fluid state. (Reprinted
with permission from Ref. 50.   Copyright © 1984 Friedr.
Vieweg & Sohn, Wiesbaden, Federal Republic of Germany.)

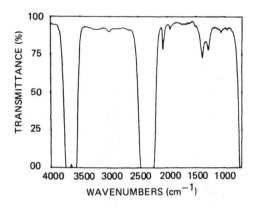

(c)

Figure 3-22 (Continued)

pressure, eluent solubility (k') can be altered. SFC pressure
programming is analogous to temperature programming in GC
and solvent programming in HPLC. Unfortunately, significant
changes in CO_2 spectra occur as pressure is increased. The
Fermi resonance bands occur in the fingerprint region of
typical organic eluents (Figure 3-24). The absorptivity of
these bands increases dramatically as pressure is increased.
When SFC pressure programming is employed, reference CO_2

Table 3-2 Some Properties of Infrared
Transparent Materials

Material	Wavelength range (cm^{-1})	M_R (psi)
NaCl	5000– 625	570
KBr	5000– 385	480
ZnSe	5000– 550	8000
ZnS	5000– 690	14,100
CaF$_2$	5000– 1110	5300

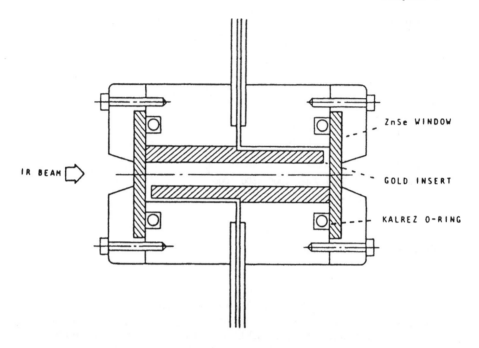

Figure 3-23   SFC/FT-IR flow cell based on a light pipe configuration. (Reproduced from Ref. 53 by permission of Preston
Publications, a division of Preston Industries, Inc., Niles, Illinois.)

spectra must be measured at each pressure that elutions occur
so that Fermi resonance bands can be eliminated by absorbance
subtraction.   Olesik et al. have found that the change in
Fermi resonance band absorbance as a function of pressure is
much less for liquid phase [Figure 3-25(a)] than supercritical
fluid [Figure 3-25(b)] detection.   They recommend that flow
cell detection of liquid CO_2 be employed rather than supercritical fluid mobile phase detection [50].
     Variation of Fermi resonance band absorption with pressure
restricts the use of flow cell measurements in the fingerprint
region to a pressure range determined by the flow cell path
length.   In order to use absorbance subtraction as a means of
removing mobile phase spectral components from eluent spectra,
mobile phase absorbance should be confined to 0.1—1 A.U.

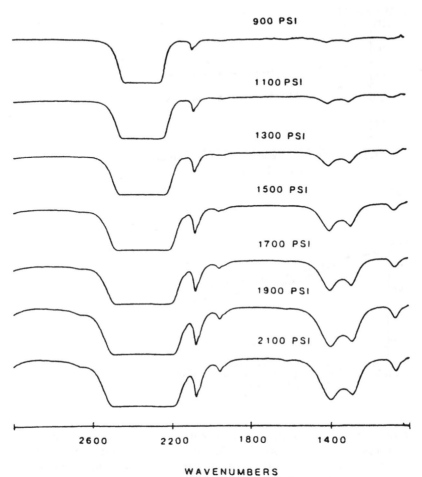

900 PSI

1100 PSI

1300 PSI

1500 PSI

1700 PSI

1900 PSI

2100 PSI

2600   2200   1800   1400

WAVENUMBERS

Figure 3-24   Variations in CO_2 transmittance spectra at different SFC column pressures. (Reprinted with permission from Ref. 15. Copyright © 1983 American Chemical Society, Washington, D.C.)

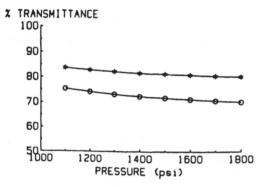

(a)

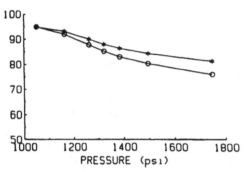

(b)

Figure 3-25   Plots of CO_2 transmittance at 1284 cm^{-1} (stars) and 1384 cm^{-1} (circles) as a function of SFC column pressure for (a) liquid phase CO_2 measurements and (b) supercritical fluid CO_2 measurements.   (Reprinted with permission from Ref. 50.   Copyright © 1984 Friedr. Vieweg & Sohn, Wiesbaden, Federal Republic of Germany.)

Because Fermi resonance band absorbance increases with increasing pressure, a maximum operating pressure exists above which the absorbance of these bands will exceed the 1 A.U. limit. Because absorbance is directly proportional to path length, the maximum operating pressure depends on cell path length. Therefore, a pressure vs path length tradeoff exists in flow cell SFC /FT-IR in addition to the elution volume vs path length tradeoff common to flow cell HPLC /FT-IR. By increasing flow cell path length, the maximum operating pressure of SFC / FT-IR flow cells is reduced. Hughes and Fasching determined that their 8 mm path length cell could not be operated at pressures above 2450 psi CO_2 (0.74 g/cm^2) at 50°C without sacrificing information in the infrared fingerprint region [52].

B. Mobile Phase Elimination Interface

Buffer memory and diffuse reflectance mobile phase elimination HPLC /FT-IR interfaces have been adapted to SFC /FT-IR [16,47,55]. The buffer memory SFC /FT-IR interface functions in the same manner as the previously described buffer memory HPLC /FT-IR interface. In the SFC /DRIFTS interface, a restrictor placed at the exit of the SFC column is used to direct column effluent onto KCl powder. SFC mobile phase (CO_2) evaporates quickly, preventing eluent diffusion. For this reason, sample cups are not required for the SFC /FT-IR interface. Instead, a KCl strip 4 mm wide, 2 mm deep, and 250 mm long has been employed [16]. The KCl strip is slowly moved during separation to deposit isolated chromatographic eluents at different positions on the strip. Detection limits in the low nanogram range have been reported for this interface.

A variation of the buffer memory interface in which an infrared microscope is employed for analysis was described by Pentoney et al. [56,57]. SFC /FT-IR eluents were deposited on a ZnSe window and later analyzed by infrared microscopy. The interface was evaluated with capillary column SFC and found to provide detection limits in the low nanogram range. The ZnSe window is moved slowly (4 μm/sec) during SFC separation to deposit the chromatogram. Profiles of eluent depositions indicate that peak half-widths of about 100 μm can be attained.

C. Matrix Isolation Interface

Raymer et al. demonstrated that the same apparatus used for GC /MI /FT-IR could also be used for SFC /MI /FT-IR [58].

A cryogenic deposition surface was maintained at 150°K instead of the usual 10°K needed for GC/MI/FT-IR measurements. At this temperature, the CO_2 mobile phase did not condense on the deposition surface and was removed by evacuation. The matrix material employed for SFC/MI/FT-IR measurements was CCl_4. Carbon tetrachloride is highly transparent in the mid-infrared range and is relatively inert. A small amount of CCl_4 was added to the CO_2 mobile phase. Detection limits below 100 ng were reported for this interface. SFC/MI/FT-IR spectra exhibit sharp absorbance bands characteristic of isolated molecules. Spectra often contain a sharp band at 2350 cm^{-1} corresponding to CO_2 mobile phase that was captured in the CCl_4 matrix during deposition. Fortunately, this band does not obscure typical organic infrared absorbance bands.

VII. SUMMARY

A general-purpose LC/FT-IR interface does not yet exist. Mobile phase interferences pose problems for both flow cell and mobile phase elimination interfaces. Table 3-3 summarizes the characteristics of LC/FT-IR interfaces developed to date. Flow cell interfaces provide no enrichment, but all eluent passes into the infrared analyzer, resulting in 100% yield. For most applications, flow cell path length must be kept below 1 mm to facilitate mobile phase spectral subtraction. This means that flow cell volume is typically one-tenth of $V_{P\frac{1}{2}}$ or less. If we assume transfer tubing volume is small, resolution degradation for flow cell interfaces is negligible. In contrast, resolution degradation for mobile phase elimination interfaces may be significant. Chromatographic resolution for solvent removal interfaces is determined by eluent diffusion during deposition and the infrared beam size employed for detection. For microbore HPLC/FT-IR, 0.5 mm diameter infrared beams must be employed to obtain adequate chromatogram temporal digitization. The main advantage of mobile phase elimination techniques derives from high enrichment factors attained. However, 100% yield can be obtained only for nonvolatile eluents. Mobile phase elimination interfaces that incorporate extraction are inherently less than 100% efficient due to distribution of eluent between mobile phase and extraction solvent. In addition, mobile phase elimination interfaces tend to concentrate solvent impurities along with separated eluents. For this reason, high-purity solvents are required for mobile phase elimination

Table 3-3   LC /FT-IR Interface Characteristics

Interface	R_d	Enrichment factor	% Yield
Transmission flow cell	∿1	1	100
ATR flow cell	∿1	1	100
Eluent deposition	>1	infinite	100[a]
Continuous reaction	>1	infinite	100[a]
Continuous extraction	>1	infinite	<100[b]

[a] Valid for nonvolatile eluents only.
[b] Dependent on eluent distribution ratio for mobile phase/ extraction solvent.

LC /FT-IR. Reported detection limits for various LC /FT-IR interfaces are given in Table 3-4. Detection limits in Table 3-4 reflect results obtained with conventional LC solvents and do not include applications of deuterated solvents. Low nanogram levels have been reported for all categories of separation except reverse phase HPLC. The ATR flow cell interface is clearly inferior to other interfaces. Microbore HPLC /FT-IR detection limits are about 10 times lower than analytical-scale HPLC /FT-IR detection limits. However, microbore column capacities are approximately 20 times lower than for analytical-scale HPLC. Thus, the useful dynamic range for microbore HPLC /FT-IR is about one-half that of analytical-scale HPLC / FT-IR. The lowest reported LC /FT-IR detection limits are approximately 100–1000 times higher than the best GC /FT-IR detection limits.

One of the most useful techniques in HPLC is solvent programming (gradient elution). Solvent programming can be used to vary eluent k' during separation. By proper selection of mobile phase constituents, solvent programming can reduce separation times and increase chromatographic resolution. Solvent programming capabilities are not easily implemented with flow cell interfaces because reference spectra for a wide range of solvent mixtures must be measured in order to effectively subtract mobile phase contributions from eluent spectra. Solvent programming represents a challenge for mobile phase

Table 3-4   LC/FT-IR Interface Detection Limits

Interface	Size exclusion	Normal phase	Reverse phase	SFC
Transmission flow cell	50 ng [27][a,b]	50 ng [27][b]	20 μg [19]	50 ng [54]
ATR flow cell	—	1 mg [32]	1 mg [32]	—
Buffer memory	1 μg [39][b]	1 μg [40][b]	—	50 ng [56]
DRIFTS	100 ng [6]	10 ng [37][b]	1 μg [35]	35 ng [55]
Reflection-absorption	—	100 ng [42][b]	—	—
Continuous reaction	—	—	100 ng [43][b]	—

[a]Estimated from normal phase detection limits in which CHCl$_3$ was the mobile phase.
[b]Microbore HPLC/FT-IR.

elimination interfaces as well. Solvent removal conditions employed at the start of a separation may not be appropriate for solvent mixtures eluting later in the solvent program. Achieving solvent programming compatibility will likely be a focus of future developments in LC /FT-IR interface design.

REFERENCES

1. R. E. Majors, *LC*, *HPLC Mag.*, *3*: 774 (1985).

2. K. L. Kizer, A. W. Mantz, and L. C. Bonar, *Am. Lab. (Fairfield, Conn.)*, *7* (5). 85 (1975).

3. D. W. Vidrine and D. R. Mattson, *Appl. Spectrosc.*, *32*: 502 (1978).

4. P. R. Griffiths, *Appl. Spectrosc.*, *31*: 497 (1977).

5. M. M. Gomez-Taylor, D. Kuehl, and P. R. Griffiths, *Int. J. Environ. Anal. Chem.*, *5*: 103 (1978).

6. D. Kuehl and P. R. Griffiths, *J. Chromatogr. Sci.*, *17*: 471 (1979).

7. N. Teramae and S. Tanaka, *Spectrosc. Lett.*, *13*: 117 (1980).

8. R. S. Brown and L. T. Taylor, *Anal. Chem.*, *55*: 1492 (1983).

9. C. C. Johnson and L. T. Taylor, *Anal. Chem.*, *55*: 436 (1983).

10. K. Jinno and C. Fujimoto, *Chromatographia*, *17*: 259 (1983).

11. P. G. Amateis and L. T. Taylor, *Anal. Chem.*, *56*: 966 (1984).

12. R. S. Brown, P. G. Amateis, and L. T. Taylor, *Chromatographia*, *18*: 396 (1984).

13. L. T. Taylor, *J. Chromatogr. Sci.*, *23*: 265 (1985).

14. C. Fujimoto, G. Uematsu, and K. Jinno, *Chromatographia*, *20*: 112 (1985).

15. K. H. Shafer and P. R. Griffiths, *Anal. Chem.*, *55*: 1939 (1983).

16. K. H. Shafer, S. L. Pentoney, Jr., and P. R. Griffiths, *Anal. Chem.*, *58*: 58 (1986).

17. D. W. Vidrine, *Fourier Transform Infrared Spectroscopy, Applications to Chemical Systems, Vol. 2* (J. R. Ferraro and L. J. Basile, eds.) Academic Press, New York, pp. 129–164 (1979).

18. B. B. Wheals, *Techniques in Liquid Chromatography* (C. F. Simpson, ed.) Wiley, New York, p. 136 (1982).

19. C. Combellas, H. Bayart, and B. Jasse, *J. Chromatogr.*, *259*: 211 (1983).

20. L. D. Rothman, S. R. Crouch, and J. D. Ingle, Jr., *Anal. Chem.*, 47: 1226 (1975).

21. D. W. Vidrine, *J. Chromatogr. Sci.*, *17*: 477 (1979).

22. R. S. Brown, D. W. Hausler, and L. T. Taylor, *Anal. Chem.*, *52*: 1511 (1980).

23. R. S. Brown, D. W. Hausler, L. T. Taylor, and R. C. Carter, *Anal. Chem.*, *53*: 197 (1981).

24. R. S. Brown and L. T. Taylor, *Anal. Chem.*, *55*: 723 (1983).

25. P. G. Amateis and L. T. Taylor, *Chromatographia, 18*: 175 (1984).

26. C. C. Johnson, J. W. Hellgeth, and L. T. Taylor, *Anal. Chem.*, *57*: 610 (1985).

27. C. C. Johnson and L. T. Taylor, *Anal. Chem.*, *56*: 2642 (1984).

28. T. Hirschfeld, *Anal. Chem.*, *50*: 1225 (1978).

29. P. K. Dasgupta, *Anal. Chem.*, *56*: 1401 (1984).

30. N. J. Harrick, *Internal Reflection Spectroscopy*, Harrick Scientific Corp., Ossining, New York, p. 30 (1979).

31. A. Rein and P. Wilks, *Am. Lab. (Fairfield, Conn.)*, *14*: (10), 197 (1982).

32. M. Sabo, J. Gross, J. Wang, and I. E. Rosenberg, *Anal. Chem.*, *57*: 1822 (1985).

33. D. T. Kuehl and P. R. Griffiths, *Anal. Chem.*, *52*: 1394 (1980).

34. M. P. Fuller and P. R. Griffiths, *Appl. Spectrosc.*, *34*: 533 (1980).

35. C. M. Conroy, P. R. Griffiths, P. J. Duff, and
    L. V. Azarraga, *Anal. Chem.*, *56*: 2636 (1984).

36. B. L. Karger, D. P. Kirby, P. Vouros, R. L. Foltz, and
    B. Hidy, *Anal. Chem.*, *51*: 2324 (1979).

37. C. M. Conroy, P. R. Griffiths, and K. Jinno, *Anal.
    Chem.*, *57*: 822 (1985).

38. K. Jinno and C. Fujimoto, *J. High Res. Chromatogr.
    Chromatogr. Commun.*, *4*: 532 (1981).

39. K. Jinno, C. Fujimoto, and D. Ishii, *J. Chromatogr.*,
    239: 625 (1982).

40. K. Jinno, C. Fujimoto, and Y. Hirata, *Appl. Spectrosc.*,
    36: 67 (1982).

41. C. Fujimoto, K. Jinno, and Y. Hirata, *J. Chromatogr.*,
    258: 81 (1983).

42. J. J. Gagel and K. Biemann, *Anal. Chem.*, *58*: 2184
    (1986).

43. K. S. Kalasinsky, J. A. S. Smith, and V. F. Kalasinsky,
    Anal. Chem., *57*: 1969 (1985).

44. R. Board, D. McManigill, and D. R. Gere, *Anal. Chem.*,
    54: 736 (1982).

45. P. A. Peaden, J. C. Fjeldsted, S. R. Springston,
    M. Novotny, and M. L. Lee, *Anal. Chem.*, *53*: 407A
    (1981).

46. T. Hirschfeld, *Anal. Chem.*, *52*: 297A (1980).

47. C. Fujimoto, Y. Hirata, and K, Jinno, *J. Chromatogr.*,
    332: 47 (1985).

48. S. B. French and M. Novotny, *Anal. Chem.*, *58*: 164
    (1986).

49. K. Ozawa, *Rev. Phys. Chem. Jpn.*, *29*: 1 (1959).

50. S. V. Olesik, S. B. French, and M. Novotny, *Chromatographia*, *18*: 489 (1984).

51. C. C. Johnson, J. W. Jordan, L. T. Taylor, and
    D. W. Vidrine, *Chromatographia*, *20*: 717 (1985).

52. M. E. Hughes and J. L. Fasching, *J. Chromatogr. Sci.*,
    23: 535 (1985).

53. J. W. Jordan and L. T. Taylor, *J. Chromatogr. Sci.*, *24*: 82 (1986).

54. R. C. Wieboldt, G. E. Adams, and D. W. Later, *Anal. Chem.*, *60*: 2422 (1988).

55. K. H. Shafer, S. L. Pentoney, and P. R. Griffiths, *J. High Res. Chromatogr. Chromatogr. Commun.*, 7: 707 (1984).

56. S. L. Pentoney, Jr., K. H. Shafer, P. R. Griffiths, and R. Fuoco, *J. High Res. Chromatogr. Chromatogr. Commun.*, 9: 168 (1986).

57. S. L. Pentoney, Jr., K. H. Shafer, and P. R. Griffiths, *J. Chromatogr. Sci.*, *24*: 230 (1986).

58. J. H. Raymer, M. A. Moseley, E. D. Pellizzari, and G. R. Vely, *J. High Res. Chromatogr. Chromatogr. Commun.*, *11*: 209 (1988).

4
Thin-Layer Chromatography/
Fourier Transform Infrared Spectroscopy

I. INTRODUCTION

Thin-layer chromatography (TLC) and high-performance thin-layer chromatography (HPTLC) are widely used for separations involving nonvolatile organic molecules. Common applications of thin-layer chromatography include the separation of pharmaceutical products, amino acids, lipids, and pesticides [1]. Compared to TLC, HPTLC provides faster separations, employs smaller-particle-size packings, has lower sample capacity, and yields lower detection limits (Table 4.1) [2]. The separating efficiency of HPTLC approaches that for nonmicrobore HPLC [3] and detection limits in the low nanogram range have been reported [2].

Silica gel is the most common stationary phase employed for TLC. However, bonded-phase packings employed for HPLC can also be used for TLC. In fact, many separations achieved by HPLC can also be performed by TLC. TLC is sometimes employed as a scout technique to quickly determine separating conditions for HPLC analysis. Isocratic HPLC capacity factor

Table 4-1   Comparison of TLC and HPTLC

Parameter	TLC	HPTLC
Sample capacity	1–5 μl	0.1 μl
Particle size	12 μm	5 μm
Typical separation duration	30–180 min	3–20 min
Detection limits	50 ng	5 ng

(k') and TLC retardation factor (R_F) can be correlated if the same stationary phase is employed for both methods [4]

$$k' = 1/R_F - 1 \qquad\qquad (4.1)$$

HPLC is usually considered the method of choice for non-volatile mixture-component separations. However, there are a number of commonly overlooked advantages of TLC for these applications. Thin-layer chromatography is simple, inexpensive, and yields high sample throughput. Unknowns and standards can be separated simultaneously on the same stationary phase under identical conditions. Even components that do not elute can be detected by TLC. In HPLC, these substances remain on the column and degrade separating efficiency. The apparatus used for TLC permits unrestricted access to the separation process. Controlled multiple solvent gradients can be applied to enhance chromatographic resolution. Physical forces such as magnetic and electric fields and thermal gradients can be employed to improve separations.

A variety of detection methods can be used for thin-layer chromatography. Separated species can be detected visually after spraying the TLC plate with a substance that reacts with components to form colored spots. Substances that fluoresce can be detected by exposing the TLC plate to ultraviolet radiation. Densitometry is used for general-purpose TLC spot detection. These methods can locate separated components but provide very little molecular structure information. As with GC and HPLC, the utility of TLC is enhanced by employing structure-specific detection methods such as mass spectrometry and infrared spectroscopy to provide eluent detection and

identification. TLC /MS interfaces have been developed for
direct analysis of separated components by using various
methods of sample vaporization [5–8]. TLC /MS suffers from
high detection limits ($\sim$1 µg), low sensitivity for polar compounds, and large background interferences. In addition,
eluent analysis is limited to substances with molecular weights
below 300, chromatographic broadening occurs due to sample
readsorption after vaporization, and changes in chromatographic
resolution occur after successive plate scans. In contrast,
TLC /FT-IR provides lower detection limits, does not damage
eluents, and does not alter chromatographic resolution.

II. BRIEF HISTORY

Early procedures for infrared analysis of thin-layer chromatography eluents involved scraping isolated components from the
chromatographic plate, extracting the eluent from stationary
phase with a suitable solvent, and dissolving the eluent in a
suitable infrared-transparent solvent for analysis by liquid cell
transmission. This process is time-consuming, especially when
numerous TLC spots are to be analyzed. Increased sample
throughput is achieved by employing faster methods for transferring TLC eluents to appropriate infrared-transparent substrates. These methods involve direct extraction of TLC spots
into infrared-transparent solvents or elution onto KBr powder
and then pressing the powder into a transmission pellet [9–12].
TLC eluent locations must be known in order to employ these
selective extraction techniques.

The first in-situ TLC /FT-IR analysis was reported by
Percival and Griffiths [13]. They constructed TLC plates
from silver chloride windows and analyzed eluents by FT-IR
transmittance with a 4X beam condenser. Detection limits of
about 1 µg were reported for this method. For silica gel TLC
plates, infrared spectra can be obtained over a frequency
range of 4000–1250 cm^{-1}. The use of alumina for TLC adsorbent
extends the usable infrared range to 1050 cm^{-1}. Refinement of
the transmittance TLC /FT-IR technique resulted in improved
detection limits and reduction of scattering effects at high frequency [14,15]. However, the method did not receive wide-
spread acceptance due to the use of nonstandard AgCl substrate for fabrication of TLC plates. Instead, a method was
sought that could be employed with commercially available pre-
coated TLC materials.

Two FT-IR sampling techniques have been developed for infrared analysis of TLC eluents in the presence of stationary phase. Diffuse reflectance FT-IR (DRIFTS) was first proposed as a TLC detection method by Fuller and Griffiths [16] and later refined by Zuber et al. [17]. Detection limits in the low μg range are obtained by this method. Photoacoustic FT-IR analysis (PAS) of TLC eluents was first demonstrated by Lloyd et al. [18]. They reported detection limits of about 90 $\mu g/cm^2$ for their apparatus. Subsequently, TLC/FT-IR/PAS sensitivity comparable to diffuse reflectance measurements was demonstrated by White [19].

The automated sample transfer interface constructed by Shafer et al. [20] represents a significant advance in TLC/ FT-IR technology. This interface can be used to simultaneously transfer all TLC eluents from highly absorbing silica or alumina substrate to infrared-transparent KCl without significant loss in chromatographic resolution. A version of this interface is currently sold by Analect Instruments under the tradename Chromalect[TM] [21,22]. Submicrogram detection limits can be obtained with this apparatus and measured spectra are similar to conventional KBr pellet transmission spectra.

III. IN SITU DIFFUSE REFLECTANCE ANALYSIS

Diffuse reflectance FT-IR (DRIFTS) is well suited to the highly scattering nature of stationary phases commonly employed for TLC. DRIFTS signal is proportional to the intensity of scattered radiation. Kubelka and Munk derived a relation between diffuse reflectance (R_∞) and absorbance (k) that is applicable to weakly absorbing materials [23,24]

$$f(R_\infty) = (1 - R_\infty)^2/2R_\infty = k/s \qquad (4.2)$$

In Eq. (4.2), R_∞ is the reflectance of an infinitely thick sample and s the scattering coefficient of the material. For the case in which the reference material is nonabsorbing,

$$k = 2.303aC \qquad (4.3)$$

where a is Beer's law absorptivity and C the concentration of the sample in the nonabsorbing matrix [25]. Figure 4-1 shows a DRIFTS spectrum in Kubelka—Munk units for 10% silica in KBr.

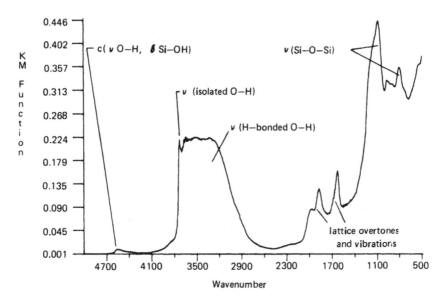

Figure 4-1   DRIFTS spectrum of 10% silica gel in KBr plotted in Kubelka—Munk units.

The silica infrared spectrum is dominated by a broad band extending from 1000 to 1200 cm^{-1}, caused by Si—O—Si stretching vibrations [26].   A band at 940 cm^{-1} produced by Si—O vibrations indicates the presence of Si—OH surface groups [27,28].   In addition, hydrogen-bonded O—H stretches are observed as a broad band extending from 3700 to 2700 cm^{-1}.

The relation derived by Kubelka and Munk [Equation (4.2)] is not applicable to highly absorbing matrices such as neat silica or alumina.   As a result, eluent difference bands obtained from in-situ TLC /DRIFTS exhibit diminished intensity in regions of high stationary phase absorbance compared to the same bands in KBr pellet transmission spectra [29].

DRIFTS spectra for some common TLC stationary phases are shown in Figure 4-2.   Silica and alumina are highly absorbing near 3400 cm^{-1}.   In addition, silica is almost totally absorbing near 1200 cm^{-1}, whereas alumina is transparent to about 1050 cm^{-1}.   Infrared spectra for silica impregnated with a fluorescent indicator are indistinguishable from spectra of silica without the indicator because the quantity of indicator

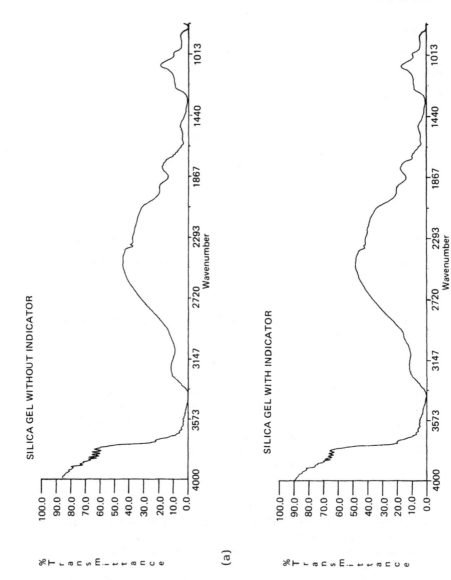

(a)

(b)

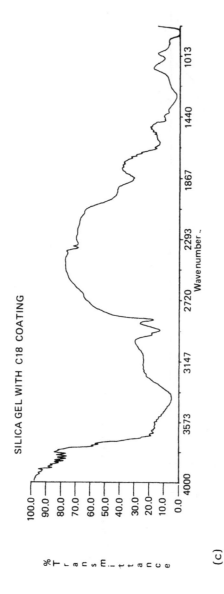

SILICA GEL WITH C18 COATING

(c)

Figure 4-2 DRIFTS spectra for common TLC stationary phases: (a) silica gel without fluorescent indicator, (b) silica gel with fluorescent indicator, (c) silica gel with C18 bonded phase, and (d) aluminum oxide with fluorescent indicator.

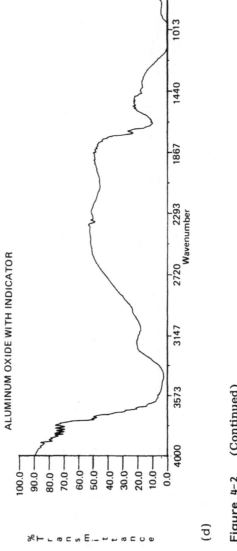

(d)

Figure 4-2 (Continued)

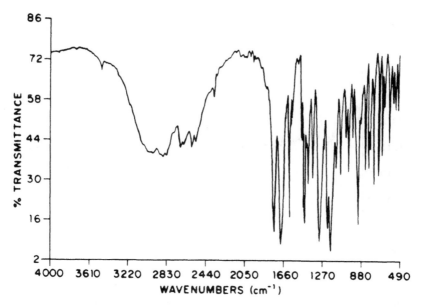

(a)

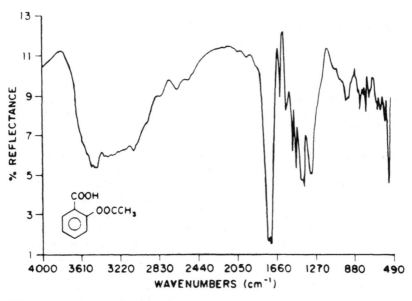

(b)

Figure 4-3   Infrared spectra for acetylsalicylic acid obtained
by (a) KBr pellet transmission and (b) in-situ TLC/DRIFTS.
(Reprinted with permission from Ref. 17.   Copyright © 1984
American Chemical Society, Washington, D.C.)

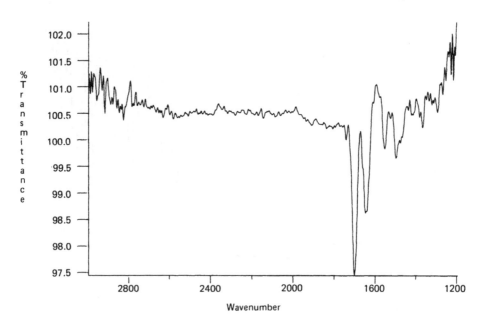

Figure 4-4   DRIFTS difference spectrum for 500 ng of caffeine deposited on aluminum oxide.

added is very small [Figures 4-2(a) and 4-2(b)]. Infrared spectra of bonded phase TLC substrates contain noticeable absorbance bands in the C—H stretching region [Figure 4-2(c)]. C_{18} bonded phase silica exhibits greater transparency than neat silica because less radiation is absorbed by C_{18} surface groups than by an equal quantity of silanol (Si—OH) moieties [Figures 4-2(a) and 4-2(c)]. Spectral information for TLC eluents in regions of high stationary phase absorbance is often lost or greatly distorted in DRIFTS difference spectra. Figure 4-3 is a comparison of KBr pellet and in-situ TLC/DRIFTS spectra for acetylsalicylic acid. Many of the features found in the fingerprint region of the KBr pellet transmission spectrum are absent from the TLC/DRIFTS spectrum. Practical in-situ TLC/DRIFTS detection limits are in the low microgram range. Figure 4-4 is a difference spectrum of 500 ng of caffeine deposited on alumina. The spectrum is distorted, but characteristic carbonyl bands are readily apparent.

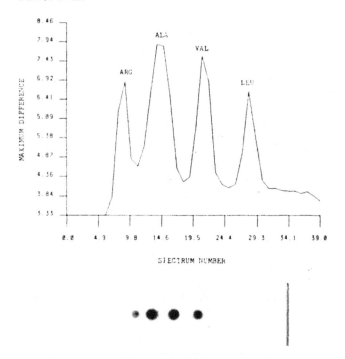

Figure 4-5   TLC /DRIFTS-reconstructed chromatogram for a
separation of arginine, alanine, valine, and leucine.   TLC
spots were developed by the addition of ninhydrin to the
chromatographic plate after separation.

One advantage of the in-situ TLC /DRIFTS interface is the
ease with which TLC plate scanning can be performed.   Commercially available DRIFTS accessories can easily be modified
to accept stepper motor-driven TLC plate translation stages.
A TLC chromatogram generated for a separation of four amino
acids on silica gel stationary phase is shown in Figure 4-5.
The chromatogram was generated by stepping the TLC plate
through the infrared beam focus of a DRIFTS accessory in
2 mm increments and computing the maximum absorbance difference [30] between spectra obtained at each position and a
reference silica spectrum measured at the beginning of the
plate scan.

IV. PHOTOACOUSTIC DETECTION

The photoacoustic phenomenon was first observed by Alexander
Graham Bell in the early 1880s. He found that when a gas
placed within a sealed compartment was illuminated with pulsed
radiation, pressure fluctuations could be detected by a sensitive microphone inside the compartment. The photoacoustic
effect has been exploited to obtain spectroscopic information
for TLC eluents in the UV-visible region by modulating incident
radiation with a chopper and detecting photoacoustic signals
generated at the chopper frequency [31,32]. In FT-IR/PAS,
incident radiation is modulated by the FT-IR interferometer
[33]. Each wavelength is modulated at a different frequency
given by

$$f = 2\bar{\nu}v \tag{4.4}$$

where f is the modulating frequency (analogous to chopping
frequency in dispersive photoacoustic spectroscopy), $\bar{\nu}$ the
wavenumber of the radiation, and v the interferometer moving-
mirror velocity. In FT-IR/PAS, the microphone detector
responds to all frequencies simultaneously. Thus, the multiplex advantage of FT-IR is retained for FT-IR/PAS analysis.
Unlike DRIFTS, in which reflected radiation is detected, FT-
IR/PAS signal is determined by the amount of radiation absorbed by the sample. For a given wavelength of radiation,
the penetration depth of infrared radiation into the sample is
dependent on the modulation frequency. Radiation modulated
at low frequency penetrates farther than radiation modulated
at higher frequencies. From Eq. (4.4), it is clear that FT-
IR/PAS spectra will exhibit the effects of longer absorption
path lengths for long infrared wavelengths compared to shorter
wavelengths. This leads to band intensity discrepancies between FT-IR/PAS spectra and KBr pellet transmission spectra
of the same material. This effect is evident in the FT-IR/PAS
spectrum of silica gel (Figure 4-6), in which low wavenumber
features are emphasized more than those at high wavenumber
compared to the corresponding DRIFTS spectrum (Figure 4-1).
Carbon black is typically employed as a reference material for
FT-IR/PAS measurements because it is highly absorbing over
the entire mid-infrared region. Sample and reference materials
must be sealed within a chamber to obtain efficient transmission
of acoustic waves through the PAS coupling gas to the microphone. For this reason, eluents must be removed from the

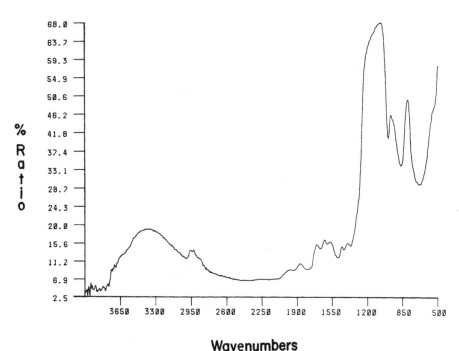

Wavenumbers

Figure 4-6   FT-IR/PAS spectrum of neat silica gel. (Reprinted
with permission from Ref. 19. Copyright © 1985 American
Chemical Society, Washington, D.C.)

TLC plate in order to be analyzed by FT-IR/PAS. This is a
major disadvantage of the TLC/FT-IR/PAS interface.

FT-IR/PAS spectra for TLC eluents are obtained by measuring the spectrum of the eluent adsorbed on the stationary
phase and subtracting spectral features corresponding to the
stationary phase. The intense Si—O—Si photoacoustic absorption (1000–1200 cm^{-1}) gives rise to a saturated response [33]
at FT-IR/PAS interferometer scan rates. As a result, infrared
difference spectra of adsorbed species contain virtually no information in this region.

Photoacoustic and KBr pellet spectra of three pharmaceutical
products are shown in Figures 4-7 through 4-9. Photoacoustic
spectra were obtained from approximately 50 μg of substance
deposited on silica gel. Spectral subtractions were used to

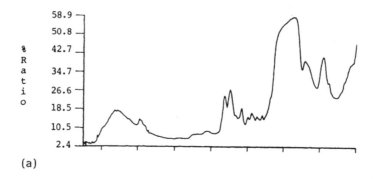

(a)

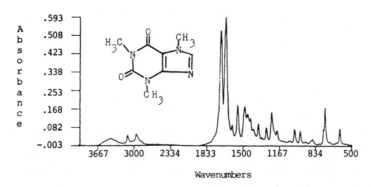

(b)

(c)

Wavenumbers

Figure 4-7 (a) Photoacoustic spectrum of 50 μg of caffeine deposited on silica gel. (b) Spectrum of adsorbed caffeine after subtraction of silica gel background. (c) Absorbance spectrum of caffeine obtained from a KBr pellet. (Reprinted with permission from Ref. 19. Copyright © 1985 American Chemical Society, Washington, D.C.)

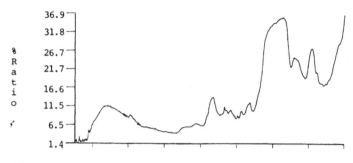

(a)

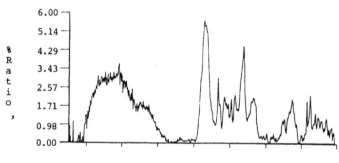

(b)

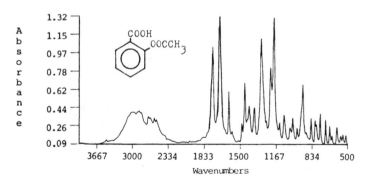

(c)

Figure 4-8   (a) Photoacoustic spectrum of 50 µg of acetylsalicylic acid deposited on silica gel.   (b) Spectrum of adsorbed
acetylsalicylic acid after subtraction of silica gel background.
(c) Absorbance spectrum of acetylsalicylic acid obtained from a
KBr pellet.   (Reprinted with permission from Ref. 19.   Copyright © 1985 American Chemical Society, Washington, D.C.)

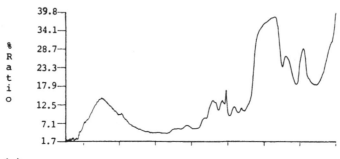

(a)

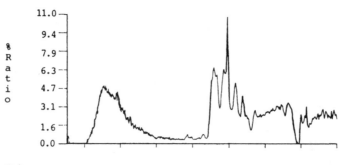

(b)

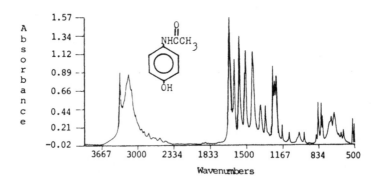

(c)

Figure 4-9 (a) Photoacoustic spectrum of 50 μg of acetoamino-phen (4-hydroxy-acetanilide) deposited on silica gel. (b) Spectrum of adsorbed acetoaminophen after subtraction of silica gel background. (c) Absorbance spectrum of acetoaminophen obtained from a KBr pellet. (Reprinted with permission from Ref. 19. Copyright © 1985 American Chemical Society, Washington, D.C.)

remove contributions from silica gel and reveal the spectrum of adsorbed species.

The photoacoustic spectrum of caffeine adsorbed on silica gel [Figure 4-7(b)] is remarkably similar to the transmittance spectrum obtained from caffeine pressed into a KBr pellet [Figure 4-7(c)]. Some band broadening is noticeable and no spectral peak information is revealed in the $1000-1200$ cm^{-1} range because of photoacoustic signal saturation.

Quantitative analysis can be obtained by TLC /FT-IR /PAS measurements for eluents present in low- to mid-microgram quantities [19]. Spectral intensity changes for the fingerprint region of caffeine TLC /FT-IR /PAS spectra are illustrated in Figure 4-10 for quantities of eluent ranging from 5 to 60 µg. A linear variation of photoacoustic signal with quantity of eluent is obtained to 20 µg. Nonlinearity at higher sample concentrations is attributed to effects from thermal saturation of the

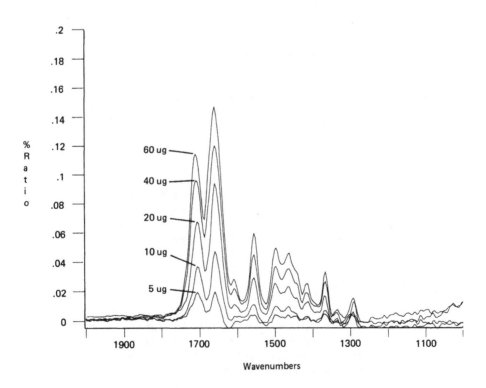

Figure 4-10   FT-IR /PAS difference spectra of caffeine for various amounts deposited on silica gel.

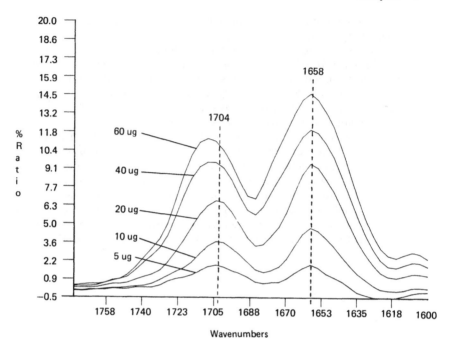

Figure 4-11   Variation of photoacoustic band location for caffeine carbonyls as a function of sample coverage. (Reprinted with permission from Ref. 19. Copyright © 1985 American Chemical Society, Washington, D.C.)

sample and has also been reported for photoacoustic measurements in the UV-visible range [31]. FT-IR/PAS detection limits for caffeine adsorbed on silica gel are estimated to be about 1 μg [19].

Inspection of the caffeine carbonyl bands at 1704 cm^{-1} and 1658 cm^{-1} reveals that the 1704 cm^{-1} band shifts to higher frequency with increasing concentration, whereas the 1658 cm^{-1} peak location does not vary (Figure 4-11). Interactions between one of the caffeine carbonyl moieties and silica gel occur when eluent coverage is low. Increased coverage reduces this interaction and the carbonyl band shifts to higher frequency. This assumption is supported by the observation that the high-frequency caffeine carbonyl stretch occurs at 1708 cm^{-1} in a nonpolar solvent such as chloroform.

Figure 4-8 contains a TLC /FT-IR /PAS spectrum for acetyl-salicylic acid obtained under conditions similar to those of the caffeine analysis. In contrast to the caffeine results, marked differences between photoacoustic and KBr pellet transmission spectra are evident. In addition to band broadening, photo-acoustic bands are red-shifted relative to KBr pellet bands. In particular, the acid carbonyl stretch is shifted 13 cm^{-1} from 1754 cm^{-1} in the KBr pellet spectrum to 1741 cm^{-1} in the photoacoustic spectrum. The shift is attributed to hydrogen bond formation with Si—OH groups on the silica gel surface.

Spectra obtained for 4-hydroxy-acetanilide (acetoaminophen) are shown in Figure 4-9. Acetoaminophen forms hydrogen bonds with the silica gel surface, as does acetylsalicylic acid. The amide carbonyl stretch is shifted from 1658 cm^{-1} in the KBr pellet transmission spectrum to 1640 cm^{-1} in the photo-acoustic spectrum. Interaction of adsorbed molecules with silica gel is extensive. As a result, the photoacoustic spectrum bears little resemblance to the KBr pellet spectrum.

V. ELUENT TRANSFER

Spectra obtained by in-situ DRIFTS and FT-IR /PAS contain band intensity distortions in regions of high stationary phase absorbance. In addition, eluent absorbance bands in TLC / FT-IR spectra shift relative to KBr pellet transmission spectra for the same material because of interactions between eluent and stationary phase. In order to avoid these problems, a sample transfer interface was designed to transfer eluents from the TLC plate to a nonabsorbing substrate prior to infrared analysis [20—22]. With this TLC /FT-IR interface, eluents are removed from a conventional TLC plate, deposited on KCl powder, and analyzed by DRIFTS. After TLC separation, the TLC plate is inserted into a metal block containing metal wicks, con-necting the TLC plate surface with a series of sample cups con-taining KCl powder (Figure 4-12). The metal strip contains 58 sample cups that are 1 mm in diameter and separated from each other by 1.6 mm. By selecting an appropriate solvent, TLC eluents can be transferred from the TLC plate to the sample cups by capillary action. An air flow across the sur-face of the sample cups evaporates the solvent and leaves eluents concentrated at the surface of the KCl substrate. Each sample cup can then be analyzed by DRIFTS (Figure 4-13). In this manner, TLC eluents can be detected without

FRONT VIEW

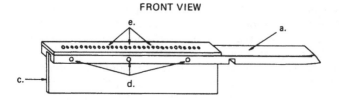

SIDE VIEW

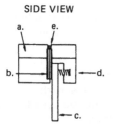

Figure 4-12   TLC/FT-IR eluent transfer apparatus.   (a) Transfer plate body, (b) metal wick, (c) TLC plate, (d) set screws,
and (e) sample cups containing KCl.   (Reprinted with permission from Ref. 22.   Copyright © 1988 International Scientific
Communications, Inc., Fairfield, Connecticut.)

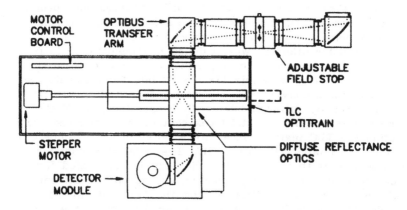

Figure 4-13   Automated DRIFTS sampler for eluents removed
from TLC plate.   (Reprinted with permission from Ref. 22.
Copyright © 1988 International Scientific Communications, Inc.,
Fairfield, Connecticut.)

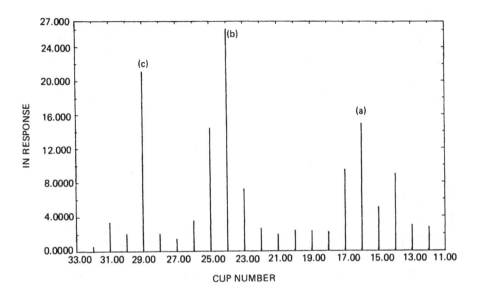

Figure 4-14   TLC /FT-IR-reconstructed chromatogram for a mixture of furosemide, oxyphenbutazone, and phenylbutazone.
(Courtesy of Analect Instruments, Inc., Irvine, California.)

interferences from the stationary phase. A portion of a TLC
chromatogram generated in this manner is shown in Figure 4-14
for a separation of three pharmaceuticals. DRIFTS spectra for
these components are shown in Figure 4-15. Each spectrum
corresponds to 2 µg of eluent. The SNR and information content of spectra obtained by eluent transfer is much greater
than that from in-situ DRIFTS or FT-IR /PAS techniques.
Figure 4-16 shows a DRIFTS spectrum of 150 ng of cocaine
separated by TLC. Clearly, quality spectra could have been
obtained for even smaller quantities of sample. The eluent
transfer TLC /DRIFTS interface is commercially available from
Analect Instruments, Inc. (Chromalect[TM]).

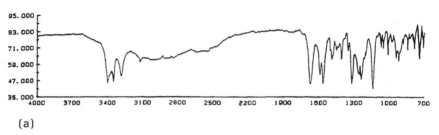

(a)

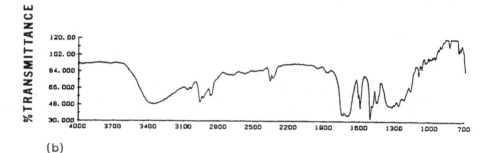

(b)

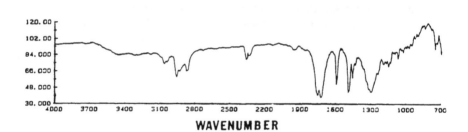

WAVENUMBER

(c)

Figure 4-15   DRIFTS spectra for (a) furosemide, (b) oxyphenbutazone, and (c) phenylbutazone.   (Courtesy of Analect Instruments, Inc., Irvine, California.)

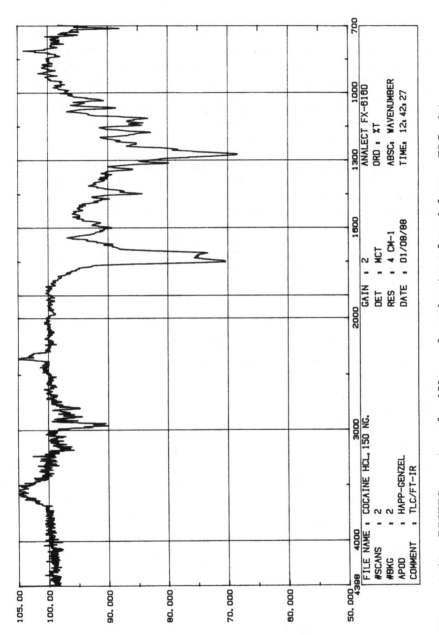

Figure 4-16 DRIFTS spectrum for 150 ng of cocaine transferred from a TLC plate. (Courtesy of Analect Instruments, Inc., Irvine, California.)

VI. SUMMARY

Lack of sensitivity and spectral distortions associated with in-situ DRIFTS and FT-IR/PAS detection methods severely restrict the utility of these techniques. Even though the sample transfer technique requires more effort to obtain infrared spectra, this effort is well worth the lower detection limits and increased spectral information gained. Resolution degradation is not a problem for in-situ detection methods and 100% yields are easily obtained when sample transfer is not required. The advantages of the sample transfer method derive from higher enrichment factors relative to the in-situ DRIFTS and FT-IR/PAS methods. Minimal resolution degradation is claimed for the eluent transfer interface [22]. However, it is not known if separated eluents are always transferred quantitatively. Thus, it is possible that yields below 100% must be tolerated for some eluents. Even with these disadvantages, the eluent transfer method is currently the method of choice for TLC/FT-IR analysis.

REFERENCES

1. J. Sherma, *Anal. Chem.*, *58*: 69R (1986).

2. H. J. Issaq, *Chromatography*, *2* (3): 37 (1987).

3. J. Sherma and B. Fried, *Anal. Chem.*, *54*: 45R (1982).

4. J. Donovan, M. Gould, and R. E. Majors, *LC-GC*, *5* (12): 1024 (1987).

5. R. Kaiser, *Chem. Br.*, *5*: 54 (1969).

6. S. E. Unber, A. Vincze, R. G. Cooks, R. Chrisman, and L. D. Rothman, *Anal. Chem.*, *53*: 976 (1981).

7. L. Ramaley, M. E. Nearing, M. A. Vaughan, R. G. Ackman, and W. D. Jamieson, *Anal. Chem.*, *55*: 2285 (1983).

8. L. Ramaley, M. A. Vaughan, and W. D. Jamieson, *Anal. Chem.*, *57*: 353 (1985).

9. R. N. McCoy and E. C. Fiebig, *Anal. Chem.*, *37*: 593 (1965).

10. D. D. Rice, *Anal. Chem.*, *39*: 1906 (1967).

11. H. R. Garner and H. Packer, *Appl. Spectrosc.*, *22*: 122 (1967).

12. W. J. deKlein, *Anal. Chem.*, *41*: 667 (1969).

13. C. J. Percival and P. R. Griffiths, *Anal. Chem.*, *47*: 154 (1975).

14. M. M. Gomez-Taylor, D. Kuehl, and P. R. Griffiths, *Appl. Spectrosc.*, *30*: 447 (1976).

15. M. M. Gomez-Taylor and P. R. Griffiths, *Appl. Spectrosc.*, *31*: 528 (1977).

16. M. P. Fuller and P. R. Griffiths, *Anal. Chem.*, *50*: 1906 (1978).

17. G. E. Zuber, R. J. Warren, P. P. Begosh, and E. L. O'Donnell, *Anal. Chem.*, *56*: 2935 (1984).

18. L. B. Lloyd, R. C. Yeates, and E. M. Eyring, *Anal. Chem.*, *54*: 549 (1982).

19. R. L. White, *Anal. Chem.*, *57*: 1819 (1985).

20. K. H. Shafer, P. R. Griffiths, and W. Shu-Qin, *Anal. Chem.*, *58*: 2708 (1986).

21. H. Bui, *Spectroscopy*, *2* (10): 44 (1987).

22. K. H. Shafer, J. A. Herman, and H. Bui, *Am. Lab. (Fairfield, Conn.)*, *20* (2): 142 (1988).

23. P. Kubelka and F. Munk, *Z. Tech. Phys.*, *12*: 593 (1931).

24. P. Kubelka, *J. Opt. Soc. Am.*, *38*: 448 (1948).

25. R. W. Frei and J. D. MacNeil, *Diffuse Reflectance Spectroscopy in Environmental Problem Solving*, CRC Press, Cleveland, Ohio (1973).

26. A. L. Smith and D. R. Anderson, *Appl. Spectrosc.*, *38*: 822 (1984).

27. M. Hino and T. Sato, *Bull. Chem. Soc. Jpn.*, *44*: 33 (1971).

28. A. Ahmed and E. Gallei, *Appl. Spectrosc.*, *28*: 430 (1974).

29. P. J. Brimmer and P. R. Griffiths, *Anal. Chem.*, *58*: 2179 (1986).

30. I. C. Bowater, R. S. Brown, J. R. Cooper, and C. L. Wilkins, *Anal. Chem.*, *58*: 2195 (1986).

31. S. L. Castleden, C. M. Elliott, G. F. Kirkbright, and D. E. M. Spillane, *Anal. Chem.*, *51*: 2152 (1979).

32. V. A. Fishman and A. J. Bard, *Anal. Chem.*, *53*: 102 (1981).

33. J. F. McClelland, *Anal. Chem.*, *55*: 89A (1983).

5
Structure Elucidation Methods

I. INTRODUCTION

Infrared absorptivity is a measure of the extent of interaction between infrared radiation and molecular motion. Relative absorptivity (relative band intensity) is dictated by the magnitude of the dipole moment change ($d\mu$) caused by the vibrational motion of the molecule (dQ). Infrared absorbance for a given molecular vibration is proportional to the square of the dipole moment change

$$A \propto \frac{d\mu^2}{dQ^2} \tag{5.1}$$

Stretching vibrations involving atoms with disparate electronegativities (e.g., OH, CO) give rise to intense infrared absorptions that are easily identified in spectra. These bands often occur at wavelengths well separated from most other organic molecule vibrations. Vibrations involving atoms with similar electronegativities (e.g., $R_1 - C - C - R_2$) give rise to low intensity absorptions and are more difficult to identify. The fingerprint region of infrared spectra comprises a large number of overlapping vibrations that represent complex molecular motions. Interpretation of infrared fingerprints is achieved by comparisons with reference spectra of compounds with similar structure.

Chromatography/FT-IR analyses produce a large number of infrared spectra. The spectroscopist may not have the time or expertise to fully interpret each spectrum. At most, verification of more easily discerned functionalities is all that can be expected. For this reason, computer-aided structure elucidation methods have been devised. These methods can be categorized as library searching, pattern recognition, and expert system techniques. The underlying premise of these methods is that a correlation exists between spectral features and molecular structure.

II. BRIEF HISTORY

Infrared spectral comparisons are frequently employed for molecular structure elucidation. Comparing reference spectra with unknowns often provides information about subtle structural features. The need for quality reference spectra stimulated efforts to compile infrared spectral data bases [1]. The first collection of infrared spectra was sponsored by the American Petroleum Institute (API) in 1943. This catalog was a compilation of infrared spectra for substances of interest to the petroleum industry. Contributions to this collection were obtained from university, government, and industry laboratories. In 1947, Sadtler Research Laboratories began to distribute a variety of infrared spectra collections. The largest infrared spectra collection to date was compiled by the American Society for Testing and Materials (ASTM). It contains approximately 140,000 spectra. Growth of this data set was terminated in the mid-1970s due to lack of funding.

The first infrared spectral libraries were simply collections of spectra contained in book form. This form is suitable for manual searching, but is incompatible with automated methods. More convenient storage was attained by transferring infrared band position information onto Hollerith punched cards (Figure 5-1). Kuentzel described the first machine-oriented library searching procedure in 1951 [2]. An IBM mechanical sorter was employed to select spectra with common features.

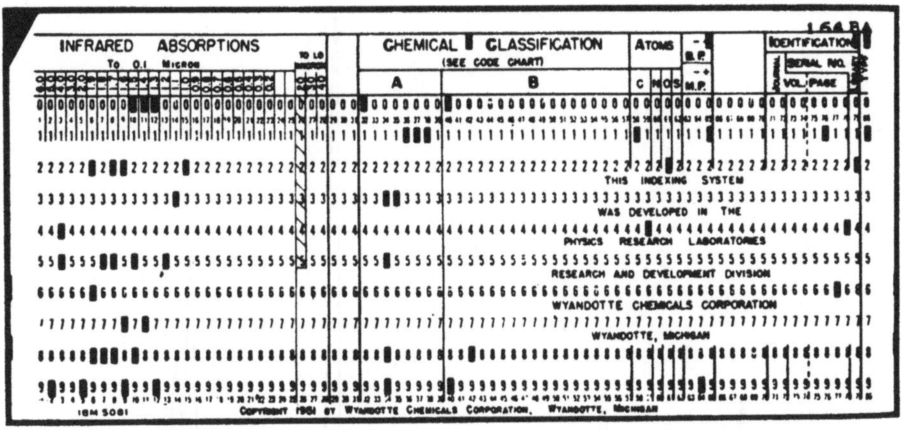

Figure 5-1 Hollerith punch card method of encoding infrared spectral information. (Reprinted with permission from Ref. 2. Copyright © 1951 American Chemical Society, Washington, D.C.)

Punched cards contained infrared absorption band positions, chemical structures, physical properties, and an index to the location of the full spectrum in hardcopy form.

Early library search applications employed binary format infrared spectra. In this form, infrared spectra were segmented into intervals and absorption bands were located by assigning a value of 1 to intervals in which a band maximum was observed and 0 to all other intervals. As shown in Figure 5-2, the resulting binary string is characteristic of the structural information contained in the spectrum. A variety of interval designations have been employed. Typically, 100 – 300 intervals are required to adequately represent mid-infrared spectra.

Pattern recognition methods were developed in the late 1960s as computer facilities became more accessible and more powerful. Most of these methods were developed and tested with the ASTM binary database. Artificial intelligence methods for spectral interpretation were introduced in the late 1970s and have been refined during the 1980s.

A 3300 spectra vapor phase GC/FT-IR infrared library was created in the late 1970s by Sadtler Laboratories for the Environmental Protection Agency (EPA). This library

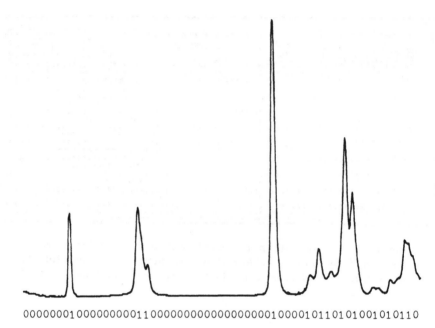

00000001 000000000011 00000000000000000000001 00001 011 01 01 001 01 0110

Figure 5-2   Binary coding for infrared absorbance bands.

consists of digitized vapor phase spectra. The digitized format contains absorbance band intensity information with sufficient accuracy that full spectra can be displayed and plotted. The EPA library became available to the general public in 1980 and provided researchers with a large quantity of infrared spectral information for testing their algorithms [3].

III. LIBRARY SEARCHING

The most common method for infrared spectral evaluation is library searching. In this method, a reference set of high-quality spectra is compared to the spectra of materials to be identified. Reference spectra are sorted in order of decreasing similarity and a list of the best matching spectra is provided to the operator. The list might contain the identity of the unknown material. More often, trends in top matches point to likely molecular functionalities present. It is important that reference spectra contained in libraries be of high quality. For this reason, guidelines have been drafted for the production of high-quality infrared spectra for use in libraries [4–8]. These guidelines describe optimum instrumental conditions for data acquisition as well as spectral data-storage formats.

Infrared library searching has proven to be a very useful technique for identifying chromatography/FT-IR mixture components. The smallest quantity of material that can be identified by chromatography/FT-IR is not the same as the detection limit for the analysis [9]. Detection limit is the smallest quantity of material for which absorbance can be observed. Identification limit is the smallest quantity of material for which structure elucidation can be achieved. Identification limits are significantly higher than detection limits. For example, GC/FT-IR detection limits range from mid-picograms to low nanograms. However, GC/FT-IR library search identification limits are typically in the high nanogram range.

Zurcher et al. developed a generalized mathematical model that summarizes all library searching algorithms [10]. Forward searching is used to find reference spectra that are most similar to the unknown spectrum. This is the most common type of search. Reverse searching is used to find all library reference spectra that may be components of an unknown spectrum. This method is commonly employed for interpretation of mixture spectra. Library searching can be used to positively identify substances only when all possible structures are represented in the library. More commonly, library searching is used to classify unknowns as to compound type.

A. Spectral Compression Algorithms

It is important that archived spectral information reflect the structures of compounds represented in the library. However, computer-storage capacities restrict the amount of spectral information that can be archived in infrared libraries. The purpose of spectral compression is to reduce the storage space required for reference spectra while retaining structure-specific information. Table 5-1 lists the characteristics of common infrared spectral compression formats. Early compression methods reduced spectra to binary format (Figure 5-2). Spectra were manually encoded from spectral plots and stored on punch cards that could be read into a computer. Band center assignments were subject to instrument calibration

Table 5-1 Common Spectral Compression Characteristics

Compression format	Information content
Binary	band center
Width-enhanced binary	band center, qualitative width
Augmented binary	band center, qualitative intensity, qualitative width
Digitized	quantitative intensity and spectral shape
Deconvoluted band	band center, quantitative intensity and width
Abstract	structure-specific eigenvalues

errors and inconsistencies in manual selection of spectral band positions. Improved infrared spectral digitization was attained by using computerized devices for translating hardcopy plots into a digital format [11,12]. Still, spectra measured by using dispersive infrared spectrometers were subject to calibration errors.

A *wiggle parameter* was employed to compensate for band center coding inconsistencies. By definition, a wiggle parameter is the number of wavelength intervals allowed for band center tolerance in binary-coded spectra comparisons. During searches, band center matches are scored when library and unknown spectra contain band maxima within this specified number of intervals. In this manner, band center locations do not have to correlate exactly in order to register a match. Another means of introducing spectrum interval tolerance for searching is by using a width-enhanced binary representation [13,14]. This is a binary storage format in which intervals on either side of band centers are assigned values of 1 to encode band width information. The interval range assigned values of 1 is selected to represent a percentage of the spectral band width. By varying this range, different amounts of band width information can be used for library searching.

Binary information can be augmented with band intensity, as well as width designations [15,16]. Augmented binary representations contain qualitative band intensity information

encoded by using numbers to indicate relative magnitude
(e.g., 1=strong, 2=medium, 3=weak). Band width can be
represented in a similar manner (e.g., 1=sharp, 2=medium,
3=broad).

Since the mid 1970s much effort has been focused on developing searching methods employing digitized spectra [17–19].
A digitized infrared spectrum is a sequence of numbers representing spectral intensities at specified points throughout the
spectrum. Typically, infrared absorbance is sampled at equal
wavelength intervals and normalized by either scaling the
largest band to unit absorbance or dividing each intensity by
the square root of the sum of the squares of all intensities
[20]. Digitized spectra can be used to accurately represent
complex spectral curves. This form of reference spectrum
storage is particularly attractive because FT-IR spectrometers
generate spectra in this format. Library search spectral comparisons are made by computing point-by-point differences
between unknown and reference spectra. This method of
spectrum correlation can yield acceptable matches even if
spectra contain severely overlapping absorbance bands. A
disadvantage of digitized spectrum library searching is that a
large amount of information must be stored in order to retain
adequate spectral detail. Table 5-2 contains storage requirements (given in bits/spectrum) for digitized infrared spectra
$(4000-500 \text{ cm}^{-1})$ at various spectral resolutions. Medium resolution is generally accepted to be 4 cm^{-1} and corresponds to a

Table 5-2 Storage Requirements for
Digitized Spectra

Resolution (cm^{-1})	Bits required/spectrum
1	56,000
2	28,000
4	14,000
8	7000
16	3500
32	1750
64	875

2 cm^{-1} digitization interval. Most chromatography/FT-IR
spectra are measured at 4 cm^{-1} resolution. However, digitized
reference spectra contained in available libraries are often
stored at reduced resolution (16 cm^{-1} or 32 cm^{-1}) in order to
minimize the amount of computer-storage media needed.
Ideally, one would like to achieve spectral compression without sacrificing spectral resolution.

This author has developed a compression algorithm that
combines features of augmented binary and digitized spectra.
The algorithm consists of the five steps shown in Figure 5-3
and has been fully automated. The input spectrum must be in
digitized format, but no resolution requirement is imposed.
Deconvolution permits separation of overlapping bands contribut-
ing to complex spectra. The peak picker locates band centers
on the wavelength axis. Curve fitting is employed to adjust
individual band parameters to best match the original spectrum.
Finally, three parameters that completely define deconvoluted
spectral bands are packed into 32 bits for storage.

The individual steps used to create a deconvoluted band
representation are illustrated for ethyl alcohol in Figure 5-4.
Figure 5-4(a) is a plot of a digitized FT-IR spectrum of ethyl

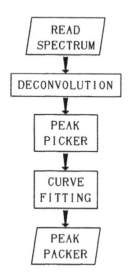

Figure 5-3   Deconvoluted band compression algorithm
flowchart.

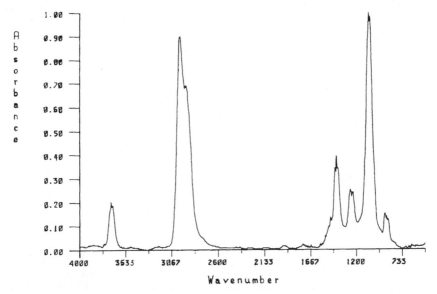

(a)

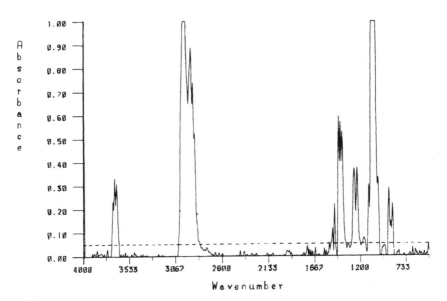

(b)

Figure 5-4 (a) GC/FT-IR vapor phase spectrum of ethanol.
(b) Result of applying deconvolution to the spectrum in (a).
(c) Component bands obtained by curve fitting. (d) Reconstructed ethanol spectrum.

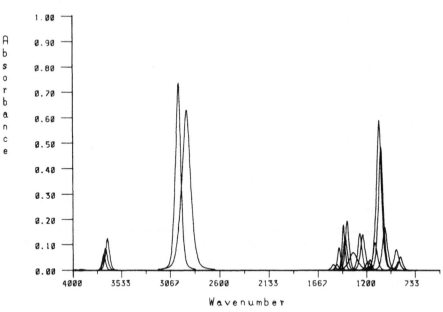

(c)

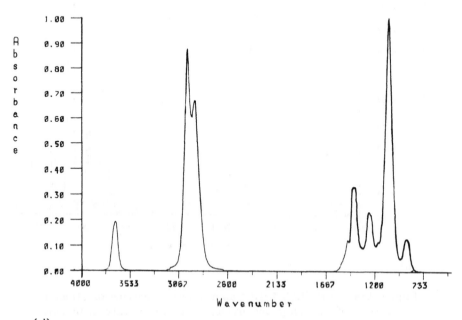

(d)

Figure 5-4   (Continued)

alcohol measured at 4 cm^{-1} resolution. Figure 5-4(b) shows the result of applying Fourier self-deconvolution to the input spectrum. The purpose of this step is to separate overlapping absorbance bands so that they will be identified by the peak picker. The deconvolution process severely distorts the original spectrum by artificially narrowing absorbance bands. The peak picker uses a threshold [dashed line in Figure 5-4(b)] to discriminate between baseline noise and absorbance bands. The peak picker determines band center locations for all bands in the spectrum and stores this information in a computer-readable file. The next step in the procedure is curve fitting. The curve fitter uses the band center locations provided by the peak picker and a user-specified band profile function to reconstruct the original spectrum. The band profile used for preliminary tests was a 1:1 mixture of Lorentzian and Gaussian shapes. The curve-fitting algorithm employs an iterative loop in which band widths and intensities are varied until the sum of all bands closely matches the original digitized spectrum. Curve fitting is achieved by minimizing the sum of the squares of differences between a digitized spectrum created by summing individual band profiles and the original digitized spectrum. Figure 5-4(c) shows the result of curve fitting for ethyl alcohol. The 24 bands derived from the compression process can be summed to obtain the spectrum shown in Figure 5-4(d). This spectrum is a close approximation of the original spectrum [Figure 5-4(a)], but was derived from information stored in 1/20th of the storage space needed for the original spectrum.

The reconstructed spectrum [Figure 5-4(d)] was derived from a set of component bands. Each band was represented by three parameters. Band intensity was stored as an 8-bit integer with an accuracy of 0.004 absorbance units (input spectra were scaled so that the largest absorbance band was unit absorbance). Band location and width were each stored in 12 bits, yielding 1 cm^{-1} accuracy. Thus, 32 bits were required to represent each band. The storage space needed to archive a compressed spectrum depends on the number of bands it contains. For typical infrared spectra, 10 − 30 deconvoluted bands can usually be isolated, requiring 320 − 960 bits for storage.

The applicability of the deconvoluted band storage format for vapor phase spectrum archival was investigated by Divis and White [21]. The procedure depicted in Figure 5-3 was applied to 3210 4 cm^{-1} vapor phase spectra. These spectra were obtained from the EPA vapor phase library after eliminating spectra with substantial baseline drift. On average, vapor phase spectra could be represented by 21.5 bands between 4000 cm^{-1} and 500 cm^{-1} requiring 688 bits for storage. This is less than the storage required for 64 cm^{-1} digitized spectra and 20 times less than the storage required for 4 cm^{-1} spectra (Table 5-2). The average difference between reconstructed (4 cm^{-1}) and original vapor phase spectra was approximately 0.02 absorbance units per digitized point ($\sim 2\%$).

A rigorous evaluation of digitized spectra reconstructed from compressed band data was made by searching each EPA vapor phase library spectrum against a database comprised of reconstructed spectra. Search lists contained the correct identity of the material in the top five matches for 99.3% of all spectra searched. The correct structure was identified as the top match in 96.5% of all searches. This indicates that reconstructed spectra derived from compressed data retain nearly all of the structural information contained in original spectra.

Library search times for digitized spectrum searches can be decreased by employing the deconvoluted band format for reference libraries. A large fraction of library searching time is attributed to spectrum retrieval from computer-storage media. Computer disk access times are typically 1000 times slower than the rate at which computer instructions can be processed [22,23]. Thus, even though the deconvoluted band storage method requires mathematical calculation to reconstruct spectra, these calculations can be accomplished faster than reading fully digitized spectra from disk storage. Enhancement in spectrum retrieval times for deconvoluted band reconstruction compared to digitized spectrum searching should increase library search rates because the time required for spectral comparisons will be the same regardless of the storage format.

The deconvoluted band compression method should be useful for archiving spectra obtained by GC/MI/FT-IR. Matrix isolation infrared spectra contain large amounts of structural information and are characterized by extremely sharp bands. This information can be preserved only if spectra are stored at relatively high resolution (e.g., $1-2$ cm^{-1}). Deconvoluted band compression can be used to represent matrix isolation spectra in $40-80$ times less storage space than required for conventional digitized spectra (Figure 5-5).

Many other compression algorithms have been developed to reduce library storage space. Abstract compression methods represent spectra in forms that are not readily converted into

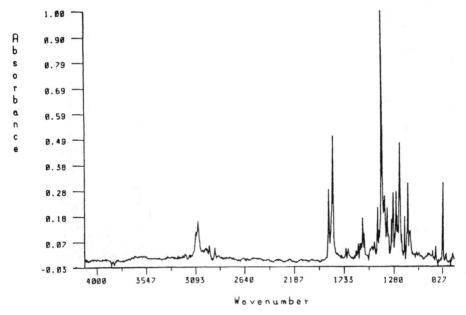

(a)

Figure 5-5 (a) GC/MI/FT-IR spectrum of cocaine. (b) Component bands obtained by curve fitting. (c) Reconstructed cocaine spectrum.

spectral information [24−27]. For example, principal component analysis has been used to isolate eigenvalues that are representative of structure. This information can be stored in much less space than spectra and can be employed for searching, but cannot be used to regenerate original spectra [25,26].

B. Search Metrics

Algorithms employed to sort library spectra by similarity to the unknown spectrum are known as *search metrics*. Library search systems based on binary spectral representations typically employ logical operations for comparisons [28−34]. Kwiatkowski and Riepe have shown that a single truth table (Table 5-3) can be used to represent all logical search

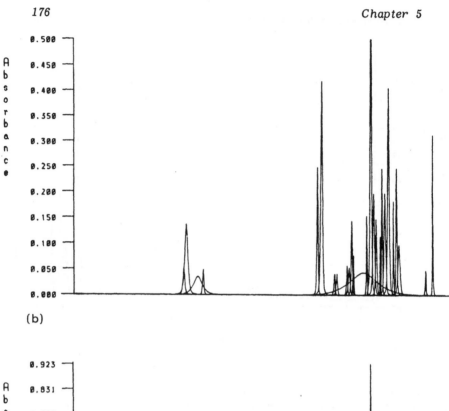

(b)

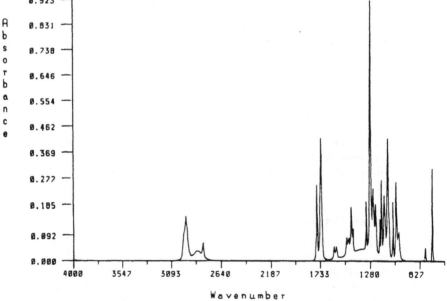

(c)

Figure 5-5 (Continued)

Table 5-3 Possible Logical Combinations of Two Logical Statements A and B

A	B	J1	J2	J3	J4	J5	J6	J7	J8	J9	J10	J11	J12	J13	J14	J15	J16
1	1	1	0	1	1	1	0	0	1	0	1	1	0	0	0	1	0
1	0	1	1	0	1	1	0	1	0	1	0	1	0	0	1	0	0
0	1	1	1	1	0	1	1	0	0	1	1	0	0	1	0	0	0
0	0	1	1	1	1	0	1	1	1	0	0	0	1	0	0	0	0

Source: Reprinted with permission from Ref. 35. Copyright © 1982 Elsevier Science Publishers, New York.

metrics [35]. Some of these operations are widely used and have been given special names (e.g., J5=logical OR, J15=logical AND), whereas others are less well known. These functions can be used to compute library similarity and dissimilarity values and can be combined in various ways to facilitate both forward and reverse searching.

Logical operators are not appropriate for augmented binary-format library comparisons. Instead, a search metric that incorporates qualitative band intensity and width information is required. One example of such a metric was devised by Penski et al. [15]

$$M = \sum_{i=1}^{n} \sum_{j=1}^{m} \exp[-(Y_j - X_i)^2/2\sigma_i] \, W_{ij} \, R_i \tag{5.2}$$

$$M_f = M \times 100.0/M_s \tag{5.3}$$

where

M is the match sum between unknown and library spectra

Y_j is the wavelength location of library spectrum band j

X_i is the wavelength location of unknown spectrum band i

σ_i is the standard deviation of unknown spectrum band i

W_{ij} is a weighting factor based on correlation between unknown and library band shapes

R_i is a weighting factor for the contribution of unknown band i to the match

M_s is the match sum of the unknown spectrum with itself

M_f is the match factor

Weighting factors (W_{ij} and R_i) are calculated from qualitative assignments of band intensity and width. Bands are ranked so that strong, sharp bands are given the maximum weighting (1.0) and weak, broad bands are given the minimum weighting (0.1).

A correlation coefficient metric was employed to evaluate the first digitized spectra searches [17]

$$M_{CORR} = 1 - \left(\frac{\sum_{i=1}^{n} U_i L_i - 1/n \left(\sum_{i=1}^{n} U_i \right) \left(\sum_{i=1}^{n} L_i \right)}{\left\{ \left[\sum_{i=1}^{n} U_i^2 - 1/n \left(\sum_{i=1}^{n} U_i \right)^2 \right] \left[\sum_{i=1}^{n} L_i^2 - 1/n \left(\sum_{i=1}^{n} L_i \right)^2 \right] \right\}^{1/2}} \right)$$

(5. 4)

where U_i and L_i are the unknown and library intensities for the ith spectral interval. Powell and Hieftje subsequently employed discrete cross-correlation functions to represent spectral matches [18]

$$M_{CC}(i\Delta n) = 1/N \sum_{n=0}^{N} U(n) \, L(n \, i\Delta n), \quad i = 0, \, 1, \, 2, \, \ldots \, N/\Delta n$$

(5. 5)

where Δn is the sampling interval in digitized spectra. Correlation-based search metrics require intensive calculations and are not appropriate for searches involving large libraries or for rapid searching typically required for chromatography/FT-IR applications. However, a modification of this method employing the odd moment of cross-correlation functions has been shown to be faster than conventional methods [36].

A simple digitized spectra search metric is the absolute-difference metric [19]

$$M_{AB} = \sum_{i=1}^{n} |U_i - L_i|$$

(5. 6)

Because digitized spectra can be considered multidimensional vectors, a Euclidean distance metric is a logical selection for spectral comparisons [19]

$$M_E = \sum_{i=1}^{n} [(U_i - L_i)^2]^{1/2}$$

(5. 7)

Computation of the square root in Eq. (5.7) does not alter the ordering of match lists. A modification of the Euclidean distance metric in which the square root is not computed is called the square-difference metric (M_{SQ}) [37]

$$M_{SQ} = \sum_{i=1}^{n} (U_i - L_i)^2 \qquad (5.8)$$

Absolute-difference and Euclidean distance (or square-difference) metrics can be computed quickly, facilitating rapid searches. These two methods are the most common comparison techniques employed for digitized spectrum searches.

Lowry and Huppler introduced two metrics based on derivative comparisons [37]

$$M_{AD} = \sum_{i=1}^{n} | \Delta U_i - \Delta L_i | \qquad (5.9)$$

$$M_{SD} = \sum_{i=1}^{n} (\Delta U_i - \Delta L_i)^2 \qquad (5.10)$$

Search results obtained by using derivative metrics are less susceptible to sloping baselines in unknown and reference spectra. This can be illustrated by comparing search results for ethanol infrared spectra with and without a sloping baseline. Figure 5-6(a) is a vapor phase spectrum of ethanol obtained by GC/FT-IR. Figure 5-6(b) is the same spectrum with a 30% baseline slope. Library search results obtained for these spectra by using the EPA vapor phase library are contained in Table 5-4. Search results for the original spectrum with the square-difference metric listed ethanol as the top match. Search results for the ethanol spectrum with 30% baseline slope did not contain ethanol in the top five matches. However, ethanol is the top match when the spectrum with baseline slope is searched with the derivative square-difference metric [Eq. (5.10)].

Availability of a wide variety of search metrics requires the spectroscopist to select the best metric for a given application. Euclidean distance (M_E or M_{SQ}) is often selected for

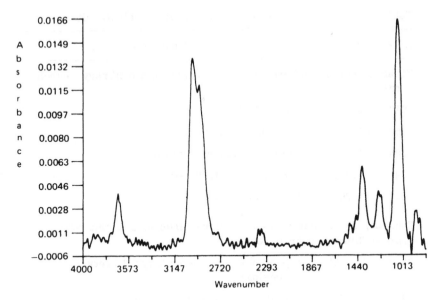

(a)

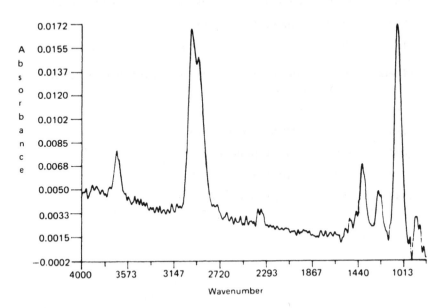

(b)

Figure 5-6 (a) GC/FT-IR vapor phase ethanol spectrum. (b) Spectrum in (a) after adding 30% baseline slope.

Table 5-4   Effects of Baseline Slope on Search Results

ID #	Index	Compound

Ethanol, square-difference metric vapor phase library search
results

519	0.1371	ethyl alcohol
456	1.0748	3-hexen-1-ol, *cis-*
385	1.0940	furfuryl alcohol, tetrahydro-
249	1.2235	1,3-butanediol, 2,2-dimethyl-
852	1.2690	1,2-propanediol

Ethanol, 30% baseline slope, square-difference metric vapor
phase library search results

2330	3.0961	1-propanol, 2-/dodecyloxy/-3-/tetra-, decyloxy/-
1989	3.1708	isobutylamine, hydrochloride
2567	3.6939	cyclohexane, 1-hexyl-4-tetradecyl-, *trans-*
2349	3.8558	dibenzotellurophene
3143	3.8625	P-xylene, A,A,A*,A*,2,3,5,6-octa-, chloro-

Ethanol, 30% baseline slope, square-derivative metric vapor
phase library search results

519	0.1106	ethyl alcohol
3018	0.3741	ethanol, 2-mercapto-
852	0.4468	1,2-propanediol
810	0.4565	1-propanol
1529	0.4681	1,3-propanediol

general-purpose searching. Match factors represent relative positions of reference and unknown spectra in a multidimensional space. One way to evaluate search metrics is to sort library members by similarity to an unknown spectrum and compare reference spectrum clustering for different search metrics. This technique is illustrated in Figure 5-7 for an infrared spectrum obtained by GC/FT-IR analysis of a flavor mixture. Figure 5-7(a) is the infrared spectrum of the unknown substance. Figure 5-7(b) shows distributions of search matches for the 3300 library spectra obtained by using four different search metrics (M_{AB}, M_{SQ}, M_{AD}, M_{SD}). A "similarity" measure was derived in order to compare search metrics. Similarity values ranged from 0 to 100% and were computed as follows:

$$\text{Similarity (i)} = \frac{\text{match}_L - \text{match}_1}{\text{match}_N - \text{match}_1} \times 100\% \qquad (5.11)$$

where match_1 was the match factor for the top search match, match_N the match factor for the worst search match, and match_L the match factor for the library spectrum under consideration.

Search metric similarity curves shown in Figure 5-7(b) are sigmoidal. Initially, similarity decreases rapidly as match factors increase. The middle of the curve exhibits a gradual decrease in similarity. This portion of the curve represents the largest fraction of library spectra. The last part of the curve drops off dramatically. These spectra are very different from the unknown spectrum. Top five search matches obtained by using the four metrics are compiled in Table 5-5. A high frequency of ester match indicated the possible presence of this functionality. Each of the search metrics produced different top matches, and match factor differences among top five matches were small. The actual identity of the unknown was not known, but it is likely that it was not represented in the library. Based on relative magnitudes of sigmoidal curve initial slope, the derivative square-difference metric (M_{SD}) provided the best discrimination between top match and subsequent matches.

Several other search metric-testing procedures have been devised to evaluate search results [38–41]. In general, these methods consist of comparing search results for a specified set of test spectra. Early test procedures evaluated the

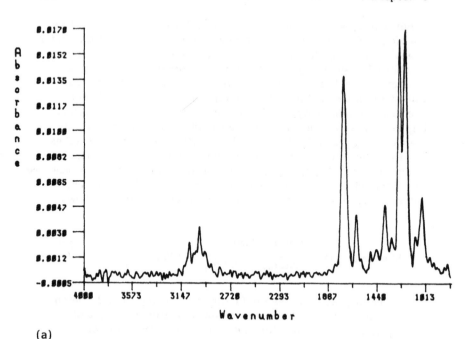

(a)

Figure 5-7   (a) GC/FT-IR vapor phase spectrum of an unknown flavor mixture component.   (b) Spectral search similarity distributions for EPA vapor phase library search of the
spectrum in (a).

ability of library searching to identify the correct substance
as the top match [38,39]. Delaney et al. developed a procedure in which metrics were compared to an arbitrary standard
search metric [40]. This method is useful even if test spectra
are not represented in the library.

C. Interferogram–Based Searching

FT-IR absorbance spectra are obtained by applying Fourier
transformation, phase correction, point-by-point division, and
logarithm computation to digitized interferograms. FT-IR interferograms contain the same spectral information as Fourier-
transformed spectra. Thus, it seems logical that interferogram

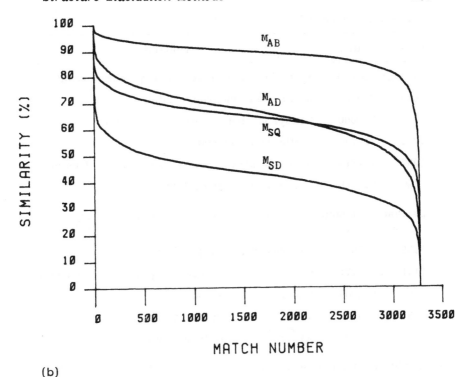

(b)

Figure 5-7   (Continued)

collections might be employed for library searching [42–44].
Unfortunately, interferograms contain phase errors and instrument functions in addition to spectral information.   These two
components are characteristic of the instrument employed and
must be removed to facilitate spectrometer-independent searching.   Phase errors are removed from FT-IR spectra by employing a phase-correction algorithm.   The algorithm developed
by Mertz is typically employed when phase correction is performed in the spectral domain [45].   An algorithm developed
by Forman et al. operates in the time (interferogram) domain
and is more appropriate for preparing interferograms for
library searching [46].   Phase-corrected interferograms still
retain an instrument function that must also be removed.
Gram–Schmidt vector subtraction can be used to remove

Table 5-5   Search Metric Comparison

ID #	Index	Compound
Absolute-difference metric vapor phase library search results		
1014	6.9765	acetic acid, isopropyl ester
1734	7.0000	benzyl alcohol, A-methyl-, acetate
1794	7.0745	acetic acid, sec-butyl ester
1957	7.2314	hexanoic acid, 2-acetyl-5-oxo-, ethyl ester
695	7.3569	acetic acid, A-methylbenzyl ester
Derivative absolute-difference metric vapor phase library search results		
1752	5.7647	cinnamic acid, ethyl ester
3141	6.0471	acrylic acid, 2-benzoyl-3-phenyl-, ethyl ester
69	6.1804	ethane, 1,2-dibromo-
56	6.4118	acetic acid, phenyl-, ethyl ester
2304	6.4824	AS-triazine-3,5/2H,4H/-dione, 2-methyl-
Euclidean-distance metric vapor phase library search results		
2247	1.0290	2-propanone, phenyl-
1041	1.0618	acetone, chloro-
1957	1.0845	hexanoic acid, 2-acetyl-5-oxo-, ethyl ester
2016	1.1629	malonic acid, isopropylidene-, di-, ethyl ester
2827	1.1678	lactamide
Derivative euclidean-distance metric vapor phase library search results		
1752	0.7932	cinnamic acid, ethyl ester
1669	0.8967	acrylic acid, ethyl ester
3141	0.9209	acrylic acid, 2-benzoyl-3-phenyl-, ethyl ester
458	0.9254	propiolic acid, ethyl ester
594	0.9255	lactic acid, 2-methyl-, ethyl ester

instrument functions from interferograms. Figure 5-8 shows
some interferograms after phase errors and instrument functions are removed. The waveforms are symmetric about zero
path difference (ZPD) and can easily be distinguished from
each other. De Haseth and Azarraga employed 100-point interferogram segments for their interferogram-based search system
[44]. Interferogram segments were treated as vectors and
the dot product between unknown and library vectors was employed as the search metric. Phase-corrected interferograms
can also be employed to classify mixture components belonging
to a given structural family [47]. This information can be
used to support library search identifications or can be employed to generate class-specific chromatograms without Fourier
transformation.

D. Real-Time GC/FT-IR Searching

The number of spectra generated by a single chromatography/
FT-IR separation can be enormous. Interpretation of this data
often requires more time than the actual separation. Data
reduction can be accelerated by using hardware array processors and co-processors. For example, a parallel-processing
GC/FT-IR data-collection/library search system featuring
simultaneous interferogram acquisition and spectral search was
developed to maximize the efficiency of the FT-IR data system
[48]. A block diagram of this data system is shown in
Figure 5-9. Two 68000 microprocessors (CPU 1 and CPU 2)
operated independently and in parallel. The processors were
linked via dual-ported, double-buffered memory and a communication register. CPU 2 was used to acquire interferogram data
and signal average in one of the two buffers (e.g., MEM A).
CPU 2 signaled CPU 1 via a communication register link when
averaging was completed. After relinquishing control of the
buffer containing averaged interferograms, CPU 2 continued to
acquire interferogram data and place it in the other buffer
(e.g., MEM B). While CPU 2 was acquiring the next data
file, CPU 1 stored the previous data on disk storage media,
computed a chromatogram intensity from the interferogram
data, and plotted this value on the data system display screen.
These functions were completed well before the next signal-
averaged interferogram was available. Prior to adding real-
time search capability, CPU 1 was idle during $50-80\%$ of the
chromatographic analysis period when $10-20$ scans were
signal-averaged for each data file (Figure 5-10). The efficiency

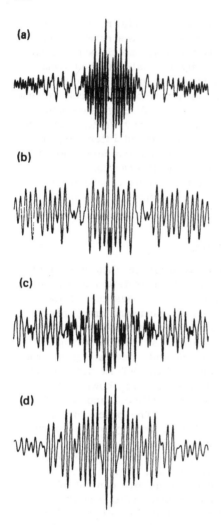

Figure 5-8 Corrected interferograms for (a) ethylbenzene, (b) *m*-dichlorobenzene, (c) phenol, and (d) acetone. (Reprinted with permission from Ref. 44. Copyright © 1981 American Chemical Society, Washington, D.C.)

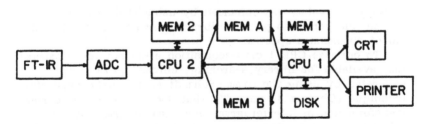

Figure 5-9   Block diagram of real-time search GC/FT-IR hardware. (Reprinted with permission from Ref. 48.   Copyright © 1987 American Chemical Society, Washington, D.C.)

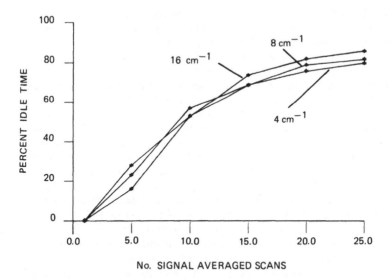

Figure 5-10   Percentage of CPU 1 data-acquisition idle time as a function of the number of signal-averaged scans per data-acquisition file.   (Reprinted with permission from Ref. 48. Copyright © 1987 American Chemical Society, Washington, D.C.)

of the data system was enhanced by using this previously wasted idle period to perform library searches for GC/FT-IR eluent absorbance spectra.

To optimize CPU 1 data-acquisition efficiency and maximize time available for library searching, chromatograms were generated by using the Gram−Schmidt vector orthogonalization technique (Chapter 1, page 23) that does not require Fourier transformation of interferograms for chromatogram intensity calculation. Chromatogram peaks were detected by a real-time peak picker that used a point-by-point slope comparison method. This method was adopted because it required minimal calculations and was insensitive to chromatogram baseline fluctuations. A flowchart for the chromatographic peak detection algorithm is depicted in Figure 5-11. Each chromatogram intensity (CURRENT) was compared with the previous value (LAST) to calculate the slope of the chromatogram. Previous slope tendencies were saved for comparison by setting software flags (POS_SLOPE, NEG_SLOPE) to true or false as appropriate. A positive slope indicated the beginning of a possible chromatographic peak elution. When a positive slope was first encountered, the chromatogram intensity at this point was saved (START) and subsequent intensity values were compared until a negative slope was detected. When this occurred, the maximum chromatogram intensity of the elution was saved (PEAK) and the next occurrence of a positive slope was sought. The next positive slope marked the end of the potential chromatographic elution. Discrimination between chromatogram baseline noise and a valid chromatographic elution was based on comparisons of peak height with a preset threshold (THRESH). If both the leading (P_THRESH) and trailing (N_THRESH) peak heights exceeded the threshold, the chromatogram peak was assumed to correspond to component elution and library searching was initiated. Data-collection software detected chromatographic elutions and tabulated computer-disk storage locations of interferograms containing eluent information. This information was accessed by the search program during GC/FT-IR data acquisition. The search program retrieved interferograms, performed necessary Fourier transformations, computed absorbance spectra, and searched spectra against the EPA vapor phase library. Because the search program was assigned low priority, it was active only while CPU 1 would otherwise be idle.

Off-line vapor phase infrared searches (including FFT and absorbance computation) required approximately 30 sec per spectrum for comparisons over a $4000-700$ cm^{-1} spectral range.

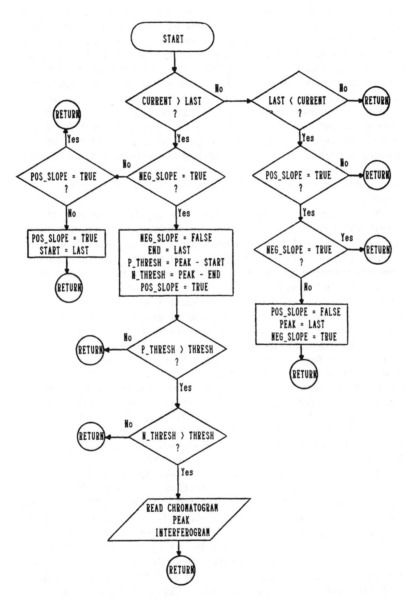

Figure 5-11 Flowchart of the algorithm used to detect GC/FT-IR chromatogram elutions during data acquisition. (Reprinted with permission from Ref. 48. Copyright © 1987 American Chemical Society, Washington, D.C.)

Real-time searches required approximately 1 min per spectrum. Real-time searches were accelerated by comparing several GC/FT-IR spectra simultaneously. About 2/3 of library searching time was found to be associated with reading the library from disk storage media. By searching several spectra at once, search times per spectrum were reduced because the library was read from disk only once. Effects of searching multiple spectra are shown graphically in Figure 5-12. When 10 spectra were searched simultaneously, the time per spectrum for real-time searching was about 36 sec per spectrum. The number of spectra searched at one time during the separation was determined by the number of chromatographic elutions that had been detected prior to each real-time GC/FT-IR search initiation.

The automated qualitative analysis capability of real-time search GC/FT-IR was evaluated by analyzing an equal volume mixture of 10 alcohols. Approximately 375 ng of each alcohol entered the GC/FT-IR light pipe. All of the selected alcohols were represented in the EPA vapor phase library. Each of the 10 chromatographic peaks shown in Figure 5-13 were detected and searched during the separation. Library search results for alcohol mixture components are compiled in Table 5-6. Seven of 10 alcohols were correctly identified as the top match. Homologues differing by one carbon atom were best matches for the other three alcohols. Search results for 1-pentanol are particularly poor. 1-Pentanol is listed as the tenth library match. Comparison of the measured 1-pentanol spectrum with the 1-pentanol library spectrum revealed significant band shape differences in the fingerprint region that accounted for the poor match. Clearly, success of automated qualitative analysis is heavily dependent on the quality of reference spectra contained in the library.

To ensure that GC/FT-IR chromatographic resolution was not degraded by adding peak detection and data-reduction capabilities, chromatograms generated with and without chromatographic peak detection were compared. Figure 5-14 contains GC/FT-IR and flame ionization (FID) chromatograms for the same naphtha mixture separated by using identical chromatographic conditions. The chromatogram in Figure 5-14(a) was generated by software incorporating real-time chromatographic peak detection and library search. Figure 5-14(b) is a chromatogram generated without chromatographic peak detection or data reduction. The temporal spacing of digitized infrared spectra in chromatogram 5-14(b) was 4.0 sec per file,

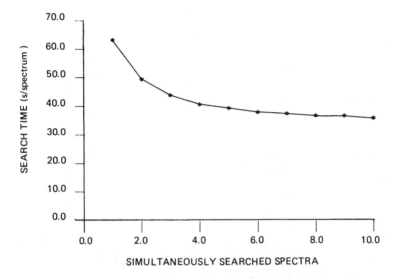

Figure 5-12   Time required for library search when multiple
spectra are searched at once.   (Reprinted with permission
from Ref. 48.   Copyright © 1987 American Chemical Society,
Washington, D.C.)

whereas the spacing in chromatogram 5-14(a) was 4.25 sec per
file.   The difference in temporal resolution (250 msec) was
attributed to operating system overhead during data acquisition
and library search process swapping.   The chromatogram in
Figure 5-14(c) is a flame ionization chromatogram (obtained
without light pipe interface) for the naphtha mixture.
A total of 47 gas chromatographic elutions were detected in
real time for the naphtha mixture.   The top three infrared
search matches for these eluents are compiled in Table 5-7.
Signal-to-noise ratio (SNR) in Table 5-7 was defined as the
ratio of the absorbance of the largest infrared band in the
spectrum to the root-mean-square (RMS) noise computed in the
$2200-2000$ cm^{-1} region of the spectrum.   The $2200-2000$ cm^{-1}
region of acquired spectra was void of absorption bands.
Match factors for top three search matches were usually compounds with similar structures.   Unequivocal component identification could not be achieved from vapor phase search results
alone.   In fact, several chromatographic eluent searches
resulted in the same top search match.   In addition, GC/FT-IR

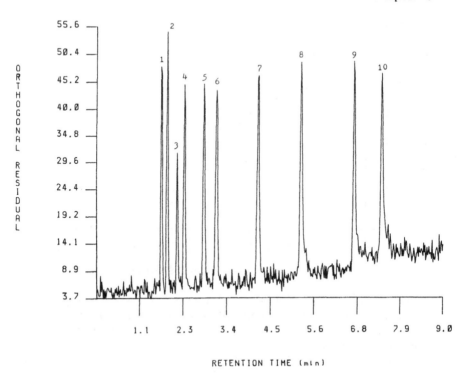

RETENTION TIME (min)

Figure 5-13   GC/FT-IR Gram–Schmidt chromatogram for a
synthetic 10-component alcohol mixture.   (Reprinted with
permission from Ref. 48.   Copyright © 1987 American Chemical
Society, Washington, D.C.)

chromatographic resolution was lower than that obtained by
using the FID detector (Figure 5-14), so several GC/FT-IR
eluent spectra undoubtedly represented component mixtures.
    GC/FT-IR eluent absorbance spectra were computed by
using a reference single-beam spectrum obtained prior to
sample injection.   Due to long-term FT-IR instability, this
reference may not have been appropriate for spectra derived
from interferograms obtained near the end of the separation.
To investigate this potential problem, GC/FT-IR eluent spectra
were extracted manually, with extra care taken to select
reference single-beam interferograms from the chromatographic
baseline near peak elutions.   Search results for manually

Table 5-6   Alcohol Mixture Search Results

Eluent	Alcohol	Top search match	Alcohol match position
1	methanol	methanol	1
2	ethanol	ethanol	1
3	t-butanol	t-butanol	1
4	1-propanol	1-propanol	1
5	iso-butanol	iso-butanol	1
6	1-butanol	1-butanol	1
7	1-pentanol	1-butanol	10
8	1-hexanol	1-hexanol	1
9	2-octanol	2-heptanol	6
10	1-octanol	1-nonanol	2

Source: Reprinted with permission from Ref. 48. Copyright ©
1987 American Chemical Society, Washington, D.C.

selected GC/FT-IR spectra are designated as "off-line" in
Table 5-7. For all 47 eluents, match factors for off-line
spectral searches were better (larger) than corresponding
real-time searches. Low SNR spectra sometimes produce
spectral search inconsistencies. Thus, spectral SNR is an
important factor and should be considered when interpreting
library search results [49].

E. Matrix Isolation Spectral Search Considerations

More absorption bands are observed in matrix isolation infrared
spectra than corresponding vapor phase spectra. Matrix
effects are primarily responsible for this observation. Matrix
perturbations can induce absorptions for vibrational modes
that are inactive in the vapor state [50]. Additional absorptions can occur due to coupling of vibrational modes with
lattice (matrix) vibrations. Trapping sites in the matrix need
not be identical. Slight changes in the environment of trapped
species can produce absorbance band multiplicity. In addition,

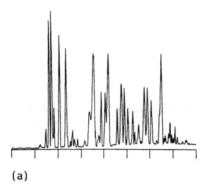

(a)

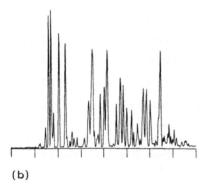

(b)

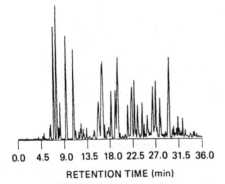

| 0.0 | 4.5 | 9.0 | 13.5 | 18.0 | 22.5 | 27.0 | 31.5 | 36.0 |

RETENTION TIME (min)

(c)

Figure 5-14   Gas chromatograms for a naphtha hydrocarbon mixture.   (a) GC/FT-IR Gram–Schmidt chromatogram obtained with chromatographic eluent detection and real-time library search; (b) GC/FT-IR Gram–Schmidt chromatogram obtained without eluent detection; (c) flame ionization chromatogram. (Reprinted with permission from Ref. 48.   Copyright © 1987 American Chemical Society, Washington, D.C.)

Table 5-7   Naphtha Mixture Search Results

No.	Ret. time (min)	SNR	Index	Real-time	Index	Off-line
1	5.53	42	0.8864	2,4-dimethyl pentane	0.9734	isooctane
			0.8863	2,5-dimethyl hexane	0.9729	2,4-dimethyl pentane
			0.8858	2,2,4-trimethyl hexane	0.9726	2,5-dimethyl hexane
2	6.60	201	0.9682	ethyl cyclohexane	0.9832	cyclohexanebutyronitrile
			0.9676	1-/5,6,7,8-tetrahydro-2-naphthyl-hexadecane	0.9820	ethyl cyclohexane
			0.9672	butyl cyclohexane	0.9813	cyclohexane
3	7.20	660	0.9826	3,3-dimethyl heptane	0.9840	3,3-dimethyl heptane
			0.9790	butyl cyclopentane	0.9809	2,2-dimethyl heptane
			0.9786	2,2-dimethyl heptane	0.9799	butyl cyclopentane
4	7.67	538	0.9806	3,3-dimethyl heptane	0.9821	3,3-dimethyl heptane
			0.9767	3-ethyl pentane	0.9775	3-ethyl pentane
			0.9764	3-methyl pentane	0.9773	2-methyl butane
5	8.14	108	0.9765	3,3-dimethyl heptane	0.9822	3,3-dimethyl heptane
			0.9732	2-methyl butane	0.9785	2,2-dimethyl heptane
			0.9732	butyl cyclopentane	0.9782	2-methyl butane

Table 5-7 (Continued)

No.	Ret. time (min)	SNR	Index	Real-time	Index	Off-line
6	9.21	565	0.9815	heptane	0.9862	heptane
			0.9773	amyl disulfide	0.9814	amyl disulfide
			0.9758	octane	0.9790	octane
7	10.42	683	0.9858	1-/5,6,7,8-tetrahydro-2-naphthyl hexadecane	0.9871	1-/5,6,7,8-tetrahydro-2-naphthyl hexadecane
			0.9849	cyclododecane	0.9866	cyclododecane
			0.9839	butyl cyclohexane	0.9856	propyl cyclohexane
8	10.69	103	0.9299	2,2-dimethyl heptane	0.9766	2,2-dimethyl heptane
			0.9292	methyl cyclopentane	0.9742	methyl cyclopentane
			0.9291	butyl cyclopentane	0.9722	2,2,5-trimethyl hexane
9	11.29	64	0.8860	butyl cyclopentane	0.9720	methyl cyclopentane
			0.8849	pentane	0.9714	2,2-dimethyl heptane
			0.8846	3,3-dimethyl heptane	0.9674	2,5-dimethyl hexane
10	11.69	86	0.9586	2,4-dimethyl pentane	0.9696	2,4-dimethyl pentane
			0.9580	2,5-dimethyl hexane	0.9688	2,5-dimethyl hexane
			0.9575	2,2,4-trimethyl hexane	0.9687	2,2,4-trimethyl hexane

11	12.03	75	0.9194	butyl cyclopentane	0.9740	2,2-dimethyl heptane
			0.9190	2,2-dimethyl heptane	0.9710	methyl cyclopentane
			0.9181	methyl cyclopentane	0.9709	butyl cyclopentane
12	12.70	90	0.9229	methyl cyclopentane	0.9707	methyl cyclopentane
			0.9222	2,2-dimethyl heptane	0.9683	2,2,5-trimethyl hexane
			0.9208	2,2,5-trimethyl hexane	0.9683	2,2-dimethyl heptane
13	14.11	61	0.9141	2-methyl butane	0.9706	2-methyl butane
			0.9140	2,5-dimethyl hexane	0.9702	2,5-dimethyl hexane
			0.9139	2,4-dimethyl pentane	0.9702	2,4-dimethyl pentane
14	14.98	327	0.9623	butyl cyclopentane	0.9783	butyl cyclopentane
			0.9594	2,6,10,14-tetramethyl-pentadecane	0.9736	2,6,10,14-tetramethyl-pentadecane
			0.9579	3,3-dimethyl heptane	0.9719	3,3-dimethyl heptane
15	15.72	470	0.9688	cis-1,3-dimethyl cyclohexane	0.9723	cis-1,3-dimethyl cyclohexane
			0.9672	1,4-dimethyl cyclohexane	0.9706	1,4-dimethyl cyclohexane
			0.9642	6,9,12-tripropyl-heptadecane	0.9673	6,9,12-tripropyl heptadecane

Table 5-7 (Continued)

No.	Ret. time (min)	SNR	Index	Real-time	Index	Off-line
16	16.12	96	0.9334	1,1-dimethyl cyclohexane	0.9818	1,1-dimethyl cyclohexane
			0.9322	6-pentyl-1-/5,6,7,8-tetrahydro-2-naphthyl/-undecane	0.9746	isobutyl cyclohexane
			0.9311	6,11-dipentyl hexadecane	0.9735	6,11-dipentyl hexadecane
17	16.86	76	0.9165	butyl cyclopentane	0.9714	2,2-dimethyl heptane
			0.9164	2,2-dimethyl heptane	0.9708	methyl cyclopentane
			0.9160	methyl cyclopentane	0.9672	butyl cyclopentane
18	17.40	366	0.9770	*trans*-1,2-dimethyl cyclohexane	0.9862	*trans*-1,2-dimethyl cyclohexane
			0.9766	*trans*-1,2-dimethyl cyclohexane	0.9857	*trans*-1,2-dimethyl cyclohexane
			0.9741	6,11-dipentyl hexadecane	0.9814	6,11-dipentyl hexadecane
19	18.13	278	0.9680	*cis*-1,3-dimethyl cyclohexane	0.9765	*cis*-1,3-dimethyl cyclohexane
			0.9674	1,4-dimethyl cyclohexane	0.9760	1,4-dimethyl cyclohexane

No.						
20	18.73	488	0.9617	6-pentyl-1-/5,6,7,8-tetrahydro-2-naphthyl/undecane	0.9680	6-pentyl-1-/5,6,7,8-tetrahydro-2-naphthyl/undecane
			0.9809	octane	0.9872	octane
			0.9798	6,9,12-tripropyl heptadecane	0.9854	6,9,12-tripropyl heptadecane
			0.9789	nonane	0.9845	nonane
21	19.87	27	0.7960	butyl cyclopentane	0.9714	2,5-dimethyl hexane
			0.7956	2,2-dimethyl heptane	0.9695	2,4-dimethyl pentane
			0.7956	3,3-dimethyl heptane	0.9688	2,2,5-trimethyl hexane
22	20.48	255	0.9650	6,9,12-tripropyl heptadecane	0.9853	6,9,12-tripropyl heptadecane
			0.9631	6,12-diethyl-9-pentyl heptadecane	0.9827	6,12-diethyl-9-pentyl heptadecane
			0.9623	octane	0.9796	octane
23	21.28	456	0.9746	1,1-dimethyl cyclohexane	0.9814	1,1-dimethyl cyclohexane
			0.9744	1,4-dimethyl cyclohexane	0.9813	isobutyl cyclohexane
			0.9738	6-pentyl-1-/5,6,7,8-tetrahydro-2-naphthyl/undecane	0.9806	1,4-dimethyl cyclohexane

Table 5-7 (Continued)

No.	Ret. time (min)	SNR	Index	Real-time	Index	Off-line
24	21.82	216	0.9740	2,6,10,14-tetramethyl pentadecane	0.9803	2,6,10,14-tetramethyl pentadecane
			0.9691	butyl cyclopentane	0.9732	butyl cyclopentane
			0.9650	2,2-dimethyl heptane	0.9705	2-tert-butyl-5-methyl cyclohexanol
25	22.48	301	0.9689	3,3-dimethyl heptane	0.9802	3,3-dimethyl heptane
			0.9674	2,4-dimethyl hexane	0.9800	2,4-dimethyl hexane
			0.9663	2,5-dimethyl hexane	0.9786	2,5-dimethyl hexane
26	23.42	144	0.9567	carvomenthene	0.9642	carvomenthene
			0.9532	2,6,10,14-tetramethyl pentadecane	0.9611	2,6,10,14-tetramethyl pentadecane
			0.9485	2-ethyl hexylamine	0.9572	2-ethyl hexylamine
27	23.82	95	0.9211	tri-2-norbornyl methanol	0.9565	tri-2-norbornyl methanol
			0.9194	1,3-dimethyl adamantane	0.9525	1,3-dimethyl adamantane
			0.9175	2,2,5-trimethyl hexane	0.9503	1-adamantane ethanol

No.						
28	24.56	178	0.9455	3,3-dimethyl heptane	0.9780	3,3-dimethyl heptane
			0.9437	2-methyl butane	0.9744	2-methyl butane
			0.9437	3-ethyl pentane	0.9741	2,4-dimethyl hexane
29	25.10	49	0.8839	heptane	0.9703	heptane
			0.8829	hexane	0.9690	6,9,12-tripropyl heptadecane
			0.8819	butyl cyclopentane	0.9689	amyl disulfide
30	25.77	295	0.9682	heptane	0.9794	heptane
			0.9652	2,6,10,14-tetramethyl pentadecane	0.9750	2,6,10,14-tetramethyl pentadecane
			0.9633	hexane	0.9732	amyl disulfide
31	26.44	277	0.9614	hexane	0.9742	hexane
			0.9594	heptane	0.9740	heptane
			0.9565	2,6,10,14-tetramethyl pentadecane	0.9716	2-ethyl hexylamine
32	27.17	250	0.9586	cis-1,3-dimethyl cyclohexane	0.9748	1,4-dimethyl cyclohexane
			0.9586	1,4-dimethyl cyclohexane	0.9743	cis-1,3-dimethyl cyclohexane
			0.9585	6,9,12-tripropyl heptadecane	0.9739	6,9,12-tripropyl heptadecane

Table 5-7 (Continued)

No.	Ret. time (min)	SNR	Index	Real-time	Index	Off-line
33	28.18	51	0.8330	hexane	0.9665	amyl disulfide
			0.8328	heptane	0.9664	6,9,12-tripropyl heptadecane
			0.8324	carvomenthene	0.9660	6,12-diethyl-9-pentyl heptadecane
34	29.18	578	0.9814	nonane	0.9868	decane
			0.9813	decane	0.9863	nonane
			0.9796	6,9,12-tripropyl heptadecane	0.9846	hendecane
35	29.58	103	0.9046	1-carbonitrile-1-cyclo-hexene	0.9793	1-carbonitrile-1-cyclo-hexene
			0.9044	chloro-cyclohexane	0.9766	chloro-cyclohexane
			0.9038	2,2-dimethyl heptane	0.9698	bromo-cyclohexane
36	29.85	57	0.8861	6,9,12-tripropyl heptadecane	0.9755	6-/5,6,7,8-tetrahydro-2-naphthyl/hexadecane
			0.8860	nonane	0.9752	myristonitrile
			0.8860	octane	0.9745	1-decanethiol

37	30.52	86	0.8922	heptane	0.9773	heptane
			0.8922	hexane	0.9762	2,6,10,14-tetramethyl pentadecane
			0.8916	2,6,10,14-tetramethyl pentadecane	0.9726	amyl disulfide
38	30.79	225	0.9540	1-/5,6,7,8-tetrahydro-2-naphthyl/hexadecane	0.9903	1-/5,6,7,8-tetrahydro-2-naphthyl/hexadecane
			0.9532	isobutyl cyclohexane	0.9891	butyl cyclohexane
			0.9530	butyl cyclohexane	0.9891	ethyl cyclohexane
39	31.06	25	0.9196	2,5-dimethyl hexane	0.9728	2,5-dimethyl hexane
			0.9194	2,2-dimethyl heptane	0.9726	2,4-dimethyl hexane
			0.9193	butyl cyclopentane	0.9717	3,3-dimethyl heptane
40	31.39	28	0.8529	butyl cyclopentane	0.9671	2,6,10,14-tetramethyl pentadecane
			0.8522	2,6,10,14-tetramethyl pentadecane	0.9636	butyl cyclopentane
			0.8520	heptane	0.9633	2,4-dimethyl hexane
41	31.79	93	0.9106	butyl cyclopentane	0.9757	butyl cyclopentane
			0.9102	2,6,10,14-tetramethyl pentadecane	0.9751	2,6,10,14-tetramethyl pentadecane
			0.9092	2,2-dimethyl heptane	0.9747	3,3-dimethyl heptane

Table 5-7   (Continued)

No.	Ret. time (min)	SNR	Index	Real-time	Index	Off-line
42	32.26	68	0.8641	butyl cyclopentane	0.9770	butyl cyclopentane
			0.8641	hexane	0.9720	heptane
			0.8637	pentane	0.9716	hexane
43	33.20	38	0.7780	hexane	0.9377	pentane
			0.7777	pentane	0.9373	hexane
			0.7776	butyl cyclopentane	0.9369	butyl cyclopentane
44	33.80	33	0.7963	hexane	0.9586	hexane
			0.7954	heptane	0.9571	2,6,10,14-tetramethyl pentadecane
			0.7952	pentane	0.9569	heptane

45	34.07	53	0.7947	heptane	0.9482	6,9,12-tripropyl heptadecane
			0.7943	6,9,12-tripropyl heptadecane	0.9469	heptane
			0.7941	hexane	0.9462	octane
46	34.47	27	0.7563	2-/dodecyloxy/-3-tetra-decyloxy/-1-propanol	0.9561	amyl disulfide
			0.7563	1,2,3,4,5,6,7,8,9,10,11, 12-dodecahydro triphenylene	0.9554	heptane
			0.7560	1,2-difluoro-1,1,2,2-tetrachloroethane	0.9537	octane
47	36.48	42	0.7664	nonane	0.9734	decane
			0.7664	6,9,12-tripropyl heptadecane	0.9722	nonane
			0.7662	octane	0.9721	6,11-dipentyl hexadecane

Source: Reprinted with permission from Ref. 48. Copyright © 1987 American Chemical Society, Washington, D.C.

absorbance bands representing conformational isomers can be observed for some substances [51,52]. For example, absorbance bands at 3547 cm^{-1} and 3565 cm^{-1} in 2,4-dichlorophenol matrix isolation infrared spectra are most likely caused by different orientations of the trapped phenolic hydrogen atom [Figure 5-15(a)]. The phenolic hydrogen can be oriented near a chlorine or an aromatic proton. Each of these orientations would be expected to affect the O—H stretching frequency differently, giving rise to two different absorption bands. In contrast, the phenolic hydrogen of 2,4,6-trichlorophenol has only one possible orientation. As a result, the O—H stretch absorbance band for 2,4,6-trichlorophenol is a single absorbance band [Figure 5-15(b)].

If the ratio of matrix atoms to trapped molecules is small, substantial interaction between trapped molecules can occur. Aggregation can produce absorbance bands characteristic of polymeric species. Band intensities for aggregate absorptions are concentration-dependent. Figure 5-16 shows carbonyl stretch absorbance bands of isobutyl methacrylate for different amounts of solute trapped in an argon matrix. The curve represented by the solid line is a matrix isolation infrared spectrum measured for 6 ng of solute deposited in the matrix. The dashed line spectrum was measured for 18 ng of solute deposited in the matrix. The dashed line spectrum exhibits a "red" shift that is consistent with increased intermolecular interaction.

The argon matrix behaves as a solvent for trapped molecules. As a result, matrix shifts can occur that are analogous to solvent shifts [50]. Stretching vibrations tend to shift to lower frequencies, whereas bending vibrations shift to higher frequencies [53]. With the exception of polar functionalities, shifts are usually small in argon matrices commonly used for GC/MI/FT-IR. However, significant absorbance band shifts are observed for hydroxyl and carbonyl stretching vibrations (Tables 5-8 and 5-9). In addition to frequency shifts, matrix effects can also perturb infrared absorbance band intensities [54].

Intermolecular interactions in the liquid and solid state can cause infrared absorbance band distortions in measured spectra. In contrast, molecules in a vapor or isolated in a low-temperature argon matrix experience minimal intermolecular interaction because individual molecules are well separated. Thus, vapor phase infrared spectra should be more similar to matrix isolation spectra than either liquid or solid phase spectra. This

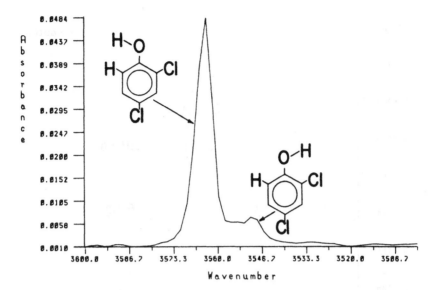

(a)

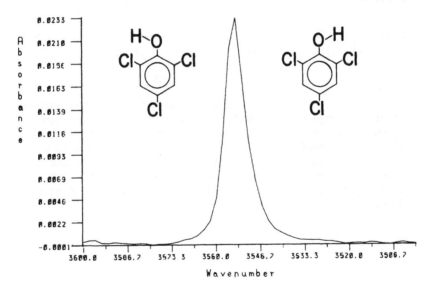

(b)

Figure 5-15 (a) Hydroxyl-stretching absorption bands of the matrix isolation infrared spectrum for 2,4-dichlorophenol. (b) Hydroxyl stretching absorption band of the matrix isolation spectrum for 2,4,6-trichlorophenol. (Reprinted with permission from Ref. 55. Copyright © 1987 Society for Applied Spectroscopy, Frederick, Maryland.)

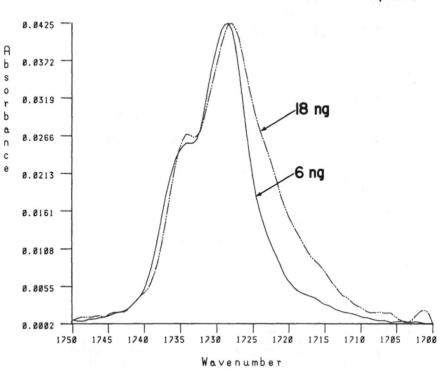

Figure 5-16 Effect of increased concentration on the carbonyl-stretching region of iso-butyl methacrylate matrix isolation spectra. Quantities indicated are amounts of solute trapped in the argon matrix. The absorbance scale corresponds to 6 ng of solute. The 18 ng absorbance band was scaled to the 6 ng band maximum. (Reprinted with permission from Ref. 55. Copyright © 1987 Society for Applied Spectroscopy, Frederick, Maryland.)

Table 5-8   Carbonyl-Stretching Vibration Matrix Shifts

| Substance | Carbonyl-stretching frequency (cm^{-1}) | | |
	Matrix-isolated	Vapor phase	Shift
Dimethyl phthalate	1741.2	1756.7	−15.5
Diethyl phthalate	1737.4	1750.9	−13.5
Dibutyl phthalate	1737.0	1749.0	−12.0
Dioctyl phthalate	1736.4	1747.0	−10.6
Ethyl laurate	1748.0	1756.7	−8.7
Butyl methacrylate	1728.2	1739.3	−11.1
Iso-butyl methacrylate	1728.9	1741.3	−12.4
Cyclohexyl acetate	1753.2	1758.6	−5.4
p-Methyl acetophenone	1696.9	1706.5	−9.6
Caffeine	1719.0	1725.8	−6.8
Citronellal	1735.9	1743.2	−7.3
Benzaldehyde	1712.2	1727.7	−15.5

Source: Reprinted with permission from Ref. 55. Copyright ©
1987 Society for Applied Spectroscopy, Frederick, Maryland.

premise was tested by broadening matrix isolation spectra by
using a smoothing algorithm and searching smoothed spectra
against vapor phase reference spectra.   When the carbonyl
stretching region was ignored for searches, it was possible to
correctly identify 28 of 40 test matrix isolation spectra as one
of the top five search matches from the EPA vapor phase
library [55]. Based on these results, routine EPA vapor
phase library searching can be expected to correctly identify
unknown matrix isolation spectra as one of the top five search
matches only 70% of the time. Much better results would be
expected if matrix isolation spectral libraries were employed.
For this reason, matrix isolation libraries are being compiled
for GC/MI/FT-IR compound identification.

Table 5-9   Hydroxyl-Stretching Vibration Matrix Shifts

Substance	Hydroxyl-stretching frequency (cm^{-1})		
	Matrix-isolated	Vapor phase	Shift
Phenol	3636.2	3649.0	-12.8
4-Nitrophenol	3626.5	3645.1	-18.6
2,6-Dimethylphenol	3636.5	3661.2	-24.7
2,4-Dichlorophenol	3562.8	3583.4	-20.6
2,4,6-Trichlorophenol	3553.2	3573.7	-20.5
o-Cresol	3633.9	3652.8	-18.9
1-Naphthol	3632.1	3652.8	-20.7
2-Naphthol	3626.4	3652.8	-26.4
Eugenol	3577.9	3593.0	-15.1

Source: Reprinted with permission from Ref. 55. Copyright ©
1987 Society for Applied Spectroscopy, Frederick, Maryland.

IV.   PATTERN RECOGNITION

The goal of pattern recognition is to develop algorithms that
are capable of recognizing the presence of molecular structure
features based solely on spectral information. Infrared spectra
are treated as multidimensional vectors or *patterns*. Each pattern (X) consists of features (x_i) representing absorbance
band locations

$$X = (x_1, \ x_2, \ x_3, \ \ldots, \ x_N) \tag{5.12}$$

Vectors containing structure-specific information often cluster
in one area of pattern space. If this clustering occurs such
that patterns belonging to one class do not overlap patterns of
other classes, it is possible to develop a discriminant function
that can be used to make class assignments. Such a discriminant function is called a binary classifier (Figure 5-17). A
collection of spectra for which class assignment is known is

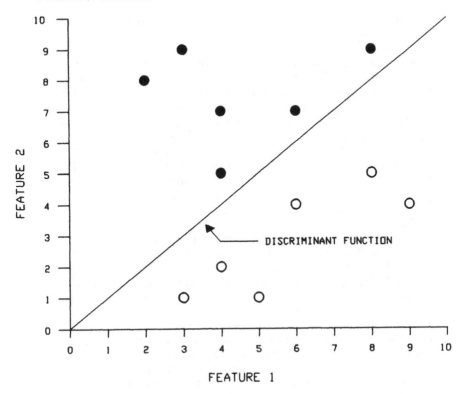

Figure 5-17   A two-dimensional pattern-recognition binary classifier. The training set is comprised of two different classes (open and filled circles).

required to develop a discriminant function. This set is called a training set. Another set of spectra for which class assignment is also known can be used to evaluate the performance of the discriminant function. This set is called a prediction set. In contrast to library searching methods, spectral collections are not needed for structure elucidation after training and predictive capabilities have been established. Classification is simply achieved by applying the discriminant function to the unknown spectrum.

Drozdov-Tichomirov described the first use of pattern recognition for infrared spectrum interpretation [56]. Carbonyl-containing spectra were classified by a potential field method. Each pattern was assumed to be a point charge in pattern space. For the training set, a positive charge was assigned to patterns belonging to the carbonyl class and a negative charge was assigned to all others. Classification of unknown spectra (patterns) was accomplished by computing the resultant field at the point at which the unknown pattern was located. A positive potential indicated that the pattern belonged to the class and a negative potential indicated that it did not belong to the class. Zero potential was obtained for patterns that were on the boundary between classes. The potential field method is computationally intensive and therefore has not been developed further for infrared spectral evaluation. The most popular methods employed for pattern recognition are the *linear learning machine* and *K-nearest-neighbor* methods.

The linear learning machine is an iterative process in which a multidimensional plane is sought that is capable of separating pattern classes. The discriminant function consists of computing the dot product of unknown patterns with a discriminant (weight) vector derived from this plane. The sign of the dot product indicates on which side of the discriminant function the unknown pattern lies and therefore the class to which it should be assigned. The numerical value of the dot product is a measure of the pattern distance from the discriminant plane. Small dot products indicate close proximity to the discriminant plane and less certain classification.

The K-nearest-neighbor (KNN) method is a cluster analysis algorithm that is used to find the closest K (where K = 1,2,3, ...) neighbors to an unknown pattern. Classification is determined by the predominate class of nearest neighbor(s). The KNN method can be employed even if complete class separation cannot be achieved. Thus, the KNN method is more tolerant to class overlap than the linear learning machine technique. However, the KNN method requires more discriminant function computations than the linear learning machine because distances between the unknown and its neighbors must be computed for each classification.

Early applications of pattern recognition to infrared spectral analysis concentrated on developing improved methods for calculating weight vectors [57−63]. Subsequent studies employed nearest-neighbor comparisons [64], and methods for reducing the number of features required for predictions were developed [65]. As expected, spectral wavelengths found to be most important for classification were often the same features contained in infrared functional group correlation charts. Thus, pattern recognition training reveals the same functional group-specific features that are learned by spectroscopists from years of experience.

Table 5-10 contains a comparison of the performance of five different pattern recognition classification methods for 2600 spectra in the 13 classes listed in Table 5-11 [66]. The maximum likelihood discriminant function was found to be the best for the spectra tested. It correctly identified class membership for 92.9% of all spectra classified. The dot product discriminant function is calculated by computing the dot product between an unknown pattern (X) and a weight vector (W)

$$D = W \cdot X \qquad\qquad (5.13)$$

Euclidean distance between the weight vector and the unknown vector is designated in Table 5-10 as the *distance* method

$$D = [(X-W) \cdot (X-W)]^{1/2} \qquad\qquad (5.14)$$

The probability that an unknown pattern belongs to a given class is represented by the maximum likelihood discriminant function [67]

$$D = \sum_{i=1}^{n} x_i \, \log\!\left(\frac{p_i}{1-p_i}\right) + \sum_{i=1}^{n} \log(1-p_i) \qquad\qquad (5.15)$$

where p_i is a conditional class probability and is represented by

$$p_i = \sum_{i=1}^{n} x_i / m \qquad\qquad (5.16)$$

In Eq. (5.16), x_i are the individual features (vector components) for the training set and m is the number of spectra belonging to the class. The last two columns in Table 5-10 represent nearest-neighbor comparisons. Hamming distance is a measure of the number of mismatches between unknown (X) and nearest-neighbor (Y) binary patterns

Table 5-10   Comparison of Pattern-Recognition Methods[a]

Class	Dot product	Distance	Maximum likelihood	Hamming distance	Tanimoto similarity
1	88.8	90.5	94.3	88.4	88.4
2	89.5	92.3	95.0	91.4	91.1
3	84.8	87.0	91.3	86.2	86.5
4	88.3	89.0	94.0	89.6	90.5
5	88.5	90.9	93.8	90.0	90.4
6	89.1	91.1	93.7	88.8	88.9
7	82.8	87.6	91.1	85.7	85.3
8	82.5	87.4	91.1	85.9	86.1
9	82.8	86.2	89.8	84.3	84.7
10	81.9	88.5	91.7	83.8	84.7
11	84.2	88.9	91.7	86.2	86.2
12	85.4	90.6	93.4	90.4	90.3
13	83.8	92.0	95.5	92.0	91.3
Average	85.6	89.5	92.9	88.0	88.1

[a]Percentage of correct assignments.
Source: Reprinted with permission from Ref. 66.   Copyright ©
1976 Society for Applied Spectroscopy, Frederick, Maryland.

$$D = \sum_{i=1}^{n} (x_i \text{ XOR } y_i) \qquad (5.16)$$

where XOR represents exclusive OR. The Tanimoto similarity
measure is a Hamming distance that has been normalized by
the number of vector components containing band centers in
either or both patterns

Table 5-11   Class Designations for
Table 5-10 Results

Class no.	Chemical functionality
1	carboxylic acid
2	ester
3	ketone
4	alcohol
5	aldehyde
6	1° amine
7	2° amine
8	3° amine
9	amide
10	urea and derivatives
11	ether and acetal
12	nitro and nitroso
13	nitrile and isonitrile

Source: Reprinted with permission
from Ref. 66. Copyright © 1976
Society for Applied Spectroscopy,
Frederick, Maryland.

$$D = \frac{\sum_{i=1}^{n}(x_i \text{ AND } y_i)}{\sum_{i=1}^{n}(x_i \text{ OR } y_i)} \tag{5.17}$$

where AND represents the logical AND operation and OR the
inclusive OR logical operation.

Reports of pattern recognition applications to infrared spectrum interpretation demonstrate the utility of these procedures by presenting results from prediction set tests. Binary spectra are often employed for training and prediction. Recently, pattern recognition has been applied to GC/FT-IR vapor phase spectral interpretations [68,69] and has been used to support proposed infrared and Raman-band assignments [70].

V. EXPERT SYSTEMS

Artificial intelligence has been applied to infrared spectral interpretation in the form of various expert systems. An expert system attempts to mimic the actions of an experienced spectroscopist in making spectrum/structure correlations [71]. Expert systems consist of a knowledge base, inference engine, and a user interface. For spectral interpretation, the knowledge base consists of interpretation rules and is referred to as procedural knowledge. The inference engine applies these rules to interpret unknown infrared spectra. The user interface is an important part of an expert system. It allows the operator freedom to modify both the knowledge base and the inference engine in ways that improve performance. Expert systems are more flexible and adaptable than pattern-recognition techniques because errors can quickly and easily be eliminated by changing the knowledge base or the inference engine. These systems can be expanded to include more class assignments by simply adding to the knowledge base [72].

Some infrared spectral interpretation expert systems are hierarchical. A series of yes/no decisions are used to assign probabilities for the presence of specific functionalities (Figure 5-18). In early systems, rules were integrated into inference engine software and not easily changed [72,73]. In 1980, Woodruff and Smith introduced the Program for Analysis of Infrared Spectra (PAIRS) that differed from previous systems in that the rules were treated as data by the inference engine [74]. PAIRS can make structure predictions for over 170 different classes (Table 5-12). A user interface was developed so that a spectroscopist could enter his/her own interpretation rules into the knowledge base. This interface was called CONCISE (Computer-Oriented Notation Concerning Infrared Spectral Evaluation) and allowed the spectroscopist to enter information without requiring a knowledge of computer programming. A block diagram of the PAIRS system is shown

Partial Aromatic Decision Tree

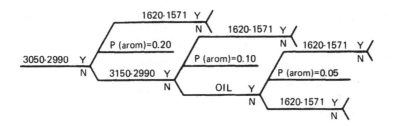

Figure 5-18 A portion of a decision tree for aromatic classification. (Reprinted with permission from Ref. 74. Copyright © 1980 American Chemical Society, Washington, D.C.)

in Figure 5-19. Arrows indicate the direction of data flow. PAIRS treats the digitized unknown spectrum, sample state, molecular formula, and interpretation rules as data and generates probabilities for the presence of each structural class in the unknown spectrum. The spectroscopist can improve structure assignments by iteratively modifying interpretation rules to eliminate obvious errors in assignments.

Spectral parameters needed by PAIRS for probability assignments are entered by the operator. Unfortunately, different operators may derive different band intensity and width information from the same spectrum, leading to discrepancies in PAIRS probability assignments. This problem is illustrated by Figure 5-20, which shows probability assignments for 136 different spectra obtained by two different operators by applying the PAIRS aldehyde-testing subroutine to these spectra. Differences in the probability distributions for the two operators are due solely to the way in which absorbance band characteristics were specified.

Several more advanced expert systems evolved from PAIRS [75−80]. The original program was adapted to run on FT-IR data systems, making it available to more users [75]. By making changes to the knowledge base and inference engine, the algorithm was adapted to vapor phase spectral applications, making it compatible with GC/FT-IR [76]. An automated method for generating interpretation rules was developed in order to

Table 5-12   PAIRS Class Designations

---

1   acetal

2   acetylene
   internal
   terminal

3   acid, carboxylic
   α-electronegative group
   pyridine-like acid
   saturated[a]
   unsaturated[b]

4   acid anhydride
   cyclic, 5-member
   cyclic, 6-member
   open chain, saturated[a]
   open chain, unsaturated[b]

5   acid halide

6   alcohol
   phenol
   primary (1°)
   secondary (2°)
   tertiary (3°)
   2° contained in ring
   3° contained in ring
   1° α-unsatn or branch[c]
   2° α-unsatn or branch[c]
   3° α-unsatn or branch[c]
   2° α,α'-unsatn and/or
     branch[c]
   3° α,α'-unsatn and/or
     branch[c]
   3° α,α',α''-unsatn and/
     or branch[c]

7   aldehyde
   saturated[a]
   unsaturated[b]

8   allene

9   amide
   acetamide
   primary
   secondary
   tertiary

10 amine
   aromatic
   primary
   secondary
   tertiary

11 aromatic
   benzene
     (monosubstituted)
   benzene (1,2)
   benzene (1,3)
   benzene (1,4)
   benzene (1,2,3)
   benzene (1,2,4)
   benzene (1,3,5)
   benzene (1,2,3,4)
   benzene (1,2,3,5)
   benzene (1,2,4,5)
   benzene (penta)
   α-naphthalene
   β-naphthalene

12 azide

13 azo

14 carbamate
   primary
   secondary
   tertiary

15 carbodiimide

16 carboxylate anion
   amino acid zwitterion
   ammonium salt

17 diacyl peroxide

18 diazo

19 ester (of carboxylic acid)
   enol[d]
   enol acetate
   lactone, 5-member, enol
   lactone, 5-member,
     saturated[a]
   lactone, 5-member,
     unsaturated[b]

Table 5-12 (Continued)

19 ester (Continued)	29 ketal
lactone, 6-member, enol	30 ketene
lactone, 6-member, saturated[a]	31 ketone
lactone, 6-member, unsaturated[b]	aryl
saturated[a,d,e]	chelate
saturated acetate[a]	cyclic, 4-member
saturated formate[a]	cyclic, 5-member
unsaturated[b]	diunsaturated[f]
unsaturated enol[b]	quinone, para
	quinone, 2 rings[g]
20 ether	saturated[a]
epoxide	unsaturated[b]
saturated[a]	β-diketone
unsaturated[b]	
	32 lactam
21 furan	4-member fused ring
	5-member fused ring
22 heteroaromatic	4-members, secondary
indole	4-members, tertiary
purine	5-members, secondary
pyrazine	5-members, tertiary
pyridine	6-members or more, secondary
pyrimidine	6-members or more, tertiary
23 hydroxylamine	
	33 mercaptan
24 imide	
cyclic, 5-member, saturated[a]	34 methyl
cyclic, 5-member, unsaturated[b]	*gem*-methyl
cyclic, 6-member, saturated[a]	35 methylene
cyclic, 6-member, unsaturated[b]	36 NH^+
open chain	37 NH_2^+
25 imine	38 NH_3^+
26 isocyanate	39 nitramine
27 isocyanide	40 nitrate
saturated[a]	41 nitrile
unsaturated[b]	saturated[a]
	unsaturated[b]
28 isothiocyanate	42 nitrite

Table 5-12   (Continued)

43	nitro α-electronegative group aromatic saturated[a] unsaturated[b]	49	sulfonate
		50	sulfone
44	olefin $CH_2=CHR$ $CH_2=CR_2$ $CHR=CR_2$ $CHR=CHR$, *cis* $CHR=CHR$, *trans*	51	sulfonic acid
		52	sulfoxide
45	oxime	53	thiocarbonyl
46	pyrrole	54	thiocyanate
47	sulfinate	55	thiophene
48	sulfonamide primary secondary tertiary	56	urea cyclic, 5-member cyclic, 6-member open chain

[a]No unsaturation α,β to the functionality.
[b]Unsaturation α,β to the functionality.
[c]Unsaturation α,β to the hydroxyl and/or branching at the α-carbon atom.
[d]Except acetate.
[e]Except formate.
[f]α,β-γ, δ or α,β-α'β' unsaturated.
[g]Quinoid carbonyls in different rings.
Source: Reprinted with permission from Ref. 74.   Copyright ©
1980 American Chemical Society, Washington, D.C.

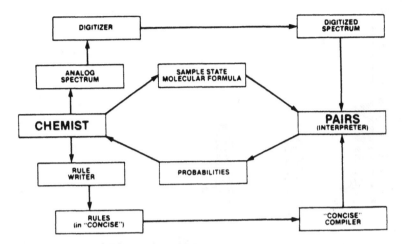

Figure 5-19 Block diagram of the PAIRS expert system.
(Reprinted with permission from Ref. 74. Copyright © 1980
American Chemical Society, Washington, D.C.)

simplify knowledge-base construction [77]. Recent versions of
PAIRS have been applied to the analysis of hazardous waste
mixtures [78-80].

Trulson and Munk constructed a table-driven expert system [81]. Probability factors and weighting constants were
assigned to spectral features that were representative of
specific functionalities. This information was stored in a table
that was used to compute class probabilities for the unknown
spectrum (Table 5-13). Saperstein designed a system in which
partial synthetic spectra were constructed, based on interpretation suggestions [82]. Partial spectra were subtracted from
the unknown spectrum to facilitate analysis of the residual
spectrum (Figure 5-21). This procedure can be repeated
until class assignments are optimized.

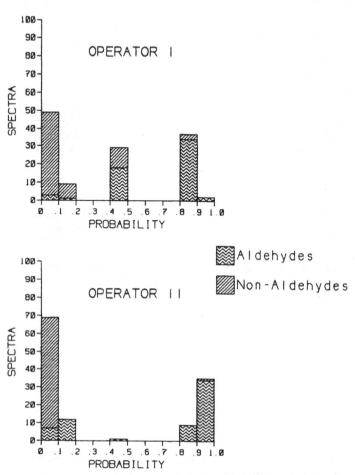

Figure 5-20   Distribution of PAIRS-derived aldehyde probabilities for 136 vapor phase spectra obtained by two different
operators.

Proposed Scheme

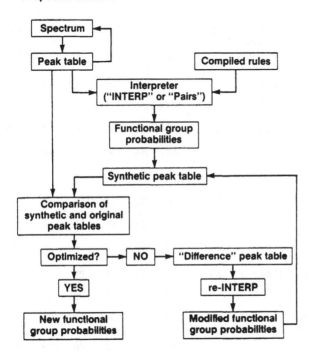

Figure 5-21   Block diagram of an optimizing expert system. (Reprinted with permission from Ref. 82. Copyright © 1986 Society for Applied Spectroscopy, Frederick, Maryland.)

Table 5-13   Chelated Ester and Aldehyde Class Parameters

Class	NCM[a]	D[b]	Range (cm⁻¹)	IW[c] (cm⁻¹)	Partial confidence values (PCV) Interval number[d]						
					1	2	3	4	5	6	7
CO_2R, chelated	20	$\nu_{C=O}$	1650–1695	5	20	20	40	60	70	80	100
		ν_{C-O}	1130–1290	20	27	70	100	75	65	50	30
					10	30	50	30	15	10	10
CHO, non-conj.	45	$\nu_{C=O}$	1705–1750	5	32	50	100	92	30	30	10
		ν_{CH}	2780–2900	20	26	80	85	100	84	32[g]	
					18	48	100	70	40	30	
					0	100	24	0	0	0	
					20	30	50	20	10	0[h]	
					30	48	76	20	0	0	
					76	24	0	0	0	0	
		ν_{CH}	2640–2780	20	20	20	20	85	100	75	66[g]
					0	22	55	85	100	75	66
					0	0	30	50	75	50	25
					24	24	38	70	100	100	88[h]
					0	10	28	80	100	48	40
					0	0	50	75	75	100	50

[a]Number of class members, i.e., number of spectra in the class.
[b]Description of diagnostic region.
[c]Width of each interval in cm⁻¹.
[d]Last column filled indicates number of intervals in diagnostic range.
[e]Weighting constant.
[f]Diagnostic region status: E = essential; NE = nonessential.
[g]Characteristic bands of alkane hydrogen absent.
[h]Characteristic bands of alkane hydrogen present.
[i]Either ν_{CH} 2780–2900 or ν_{CH} 2640–2780 must be present.
Source: Reprinted with permission from Ref. 81. Copyright © 1983 American Chemical Society, Washington, D.C.

| Partial confidence values (PCV) | | | | | | | | | | | | | | |
| Interval number[d] | | | | | | | | | | | Band descrip-tion | WC[e] | S[f] | Screens |
8	9	10	11	12	13	14	15	16	17	18				
60	20										7,8,9	0.25	E	Atom screen 0 > 2
15											8,9	0.75	E	Spectral screen ν_{OH} 3150 − 3600
10											7			cm^{-1}
10	10										7,8,9	0.30	E	
											1,2,3	0.12	i	
											4,5,6			
											7,8,9			
											1,2,3			
											4,5,6			
											7,8,9			
											1,2,3	0.58		
											4,5,6			
											7,8,9			
											1,2,3			
											4,5,6			
											7,8,9			

VI. SUMMARY

Library searching methods are the most popular chromatography/
FT-IR spectral evaluation methods available today. Search
results do not unequivocally provide the identity of unknown
species, but can indicate what functionalities are likely to be
present. Pattern-recognition methods must be trained and
therefore require a database of reference spectra. However,
after training, the database is no longer required. Artificial
intelligence methods do not require databases, but instead rely
on procedures learned by experienced spectroscopists from
numerous manual interpretations. Artificial intelligence algorithms can never be considered complete. Obvious interpretation errors are corrected by the operator as they occur.
Thus, expert system programs are in constant revision. Expert systems have promise for future applications of chroma-
tography/FT-IR. However, the greatest improvements in
chromatographic eluent structure elucidation will likely come
from combining complementary information (e.g., mass spectra,
retention indices, etc.) with infrared spectral evaluation results
to permit unequivocal structural assignments. Expert systems
that combine complementary information obtained from multiple
spectroscopic analyses are currently being developed [83–85].

REFERENCES

1. L. H. Gerantman, *Anal. Chem.*, *44* (7): 30A (1972).

2. L. E. Kuentzel, *Anal. Chem.*, *23*: 1413 (1951).

3. R. S. McDonald, *Anal. Chem.*, *52*: 361R (1980).

4. N. Fuson, R. N. Jones, H. B. Kessler, L. E. Kuentzel,
    R. S. McDonald, F. S. Mortimer, J. Overend, W. J. Potts,
    Jr., A. L. Smith, C. D. Smith, and J. E. Stewart, *Anal.
    Chem.*, *38* (9): 27A (1966).

5. P. R. Griffiths, L. V. Azarraga, J. de Haseth,
    R. W. Hannah, R. J. Jakobsen, and M. M. Ennis,
    Appl. Spectrosc., *33*: 543 (1979).

6. D. F. Gurka, M. Umana, E. D. Pellizzari, A. Moseley,
    and J. A. de Haseth, *Appl. Spectrosc.*, *39*: 297 (1985).

7. P. R. Griffiths and C. L. Wilkins, *Appl. Spectrosc.*, *42*:
    538 (1988).

8. V. F. Kalasinsky, K. G. Whitehead, R. C. Kenton, and K. S. Kalasinsky, *Appl. Spectrosc.*, *42*: 56 (1988).

9. J. R. Cooper and L. T. Taylor, *Anal. Chem.*, *56*: 1989 (1984).

10. M. Zurcher, J. T. Clerc, M. Farkas, and E. Pretsch, *Anal. Chim. Acta, 206*: 161 (1988).

11. J. A. de Haseth, W. S. Woodward, and T. L. Isenhour, *Anal. Chem.*, *48*: 1513 (1976).

12. J. T. Clerc, R. Knutti, H. Koenitzer, and J. Zupan, *Fresenius Z. Anal. Chem.*, *283*: 177 (1977).

13. F. V. Warren and M. F. Delaney, *Appl. Spectrosc.*, *37*: 172 (1983).

14. M. F. Delaney, J. R. Hallowell, Jr., and F. V. Warren, Jr., *J. Chem. Inf. Comput. Sci.*, *25*: 27 (1985).

15. E. C. Penski, D. A. Padowski, and J. B. Bouck, *Anal. Chem.*, *46*: 955 (1974).

16. R. C. Fox, *Anal. Chem.*, *48*: 717 (1976).

17. K. Tanabe and S. Saeki, *Anal. Chem.*, *47*: 118 (1975).

18. L. M. Powell and G. M. Hieftje, *Anal. Chim. Acta, 100*: 313 (1978).

19. G. T. Rasmussen and T. L. Isenhour, *Appl. Spectrosc.*, *33*. 371 (1979).

20. P. B. Harrington and T. L. Isenhour, *Appl. Spectrosc.*, *41*: 1298 (1987).

21. R. A. Divis and R. L. White, *Anal. Chem.*, *61*: 33 (1989).

22. B. Zoellick, *Byte, 11* (5): 177 (1986).

23. T. Thompson, *Byte, 13* (3): 205 (1988).

24. S. Kawata, T. Noda, and S. Minami, *Appl. Spectrosc.*, *41*: 1176 (1987).

25. C. P. Wang and T. L. Isenhour, *Appl. Spectrosc.*, *41*: 185 (1987).

26. P. B. Harrington and T. L. Isenhour, *Appl. Spectrosc.*, *41*: 449 (1987).

27. M. R. Nyden, J. E. Pallister, D. T. Sparks, and
    A. Salari, *Appl. Spectrosc.*, *41*: 63 (1987).

28. A. W. Baker, N. Wright, and A. Opler, *Anal. Chem.*, *25*:
    1457 (1953).

29. D. H. Anderson and G. L. Covert, *Anal. Chem.*, *39*:
    1288 (1967).

30. D. S. Erley, *Anal. Chem.*, *40*: 894 (1968).

31. H. B. Woodruff, S. R. Lowry, and T. L. Isenhour,
    J. Chem. Inf. Comp. Sci., *15*: 2071 (1975).

32. J. Pupan, D. Hadzi, and M. Penca, *Comp. Chem.*, *1*: 77
    (1976).

33. P. F. Dupuis and A. Dijkstra, *Fresenius Z. Anal. Chem.*,
    290: 357 (1978).

34. P. F. Dupuis, A. Dijkstra, and J. H. van der Maas,
    Fresenius Z. Anal. Chem., *291*: 27 (1978).

35. J. Kwiatkowski and W. Riepe, *Anal. Chim. Acta*, *135*: 285
    (1982).

36. J. P. Yu and H. B. Friedrich, *Appl. Spectrosc.*, *41*: 869
    (1987).

37. S. R. Lowry and D. A. Huppler, *Anal. Chem.*, *53*: 889
    (1981).

38. D. S. Erley, *Appl. Spectrosc.*, *25*: 200 (1971).

39. S. L. Grotch, *Anal. Chem.*, *45*: 2 (1973).

40. M. F. Delaney, F. V. Warren, Jr., and J. R. Hallowell,
    Jr., *Anal. Chem.*, *55*: 1925 (1983).

41. J. T. Clerc, E. Pretsch, and M. Zuercker, *Mikrochim.
    Acta*, *2*: 217 (1987).

42. G. W. Small, G. T. Rasmussen, and T. L. Isenhour,
    Appl. Spectrosc., *33*: 444 (1979).

43. L. V. Azarraga, R. R. Williams, and J. A. de Haseth,
    Appl. Spectrosc., *35*: 466 (1981).

44. J. A. de Haseth and L. V. Azarraga, *Anal. Chem.*, *53*:
    2292 (1981).

45. L. Mertz, *Infrared Phys.*, *7*: 17 (1967).

46. M. L. Forman, W. H. Steel, and G. A. Vanasse, *J. Opt. Soc. Am.*, *56*: 59 (1966).

47. P. T. Richardson and J. A. de Haseth, *Anal. Chem.*, *60*: 386 (1988).

48. R. E. Fields III and R. L. White, *Anal. Chem.*, *59*: 2709 (1987).

49. J. R. Hallowell, Jr. and M. F. Delaney, *Anal. Chem.*, *59*: 1544 (1987).

50. A. J. Barnes and H. E. Hallam, *Vibrational Spectroscopy — Modern Trends* (A. J. Barnes and W. J. Orville-Thomas, eds.) Elsevier Scientific Publishing Co., New York, p. 66 (1977).

51. A. J. Barnes, *Rev. Anal. Chem.*, *1*: 193 (1972).

52. C. J. Purnell, A. J. Barnes, S. Suzuki, D. F. Ball, and W. J. Orville-Thomas, *Chem. Phys.*, *12*: 77 (1976).

53. S. Cradock and A. J. Hinchcliffe, *Matrix Isolation*, Cambridge University Press, New York, pp. 98−99 (1975).

54. C. M. Huggins and G. C. Pimentel, *J. Chem. Phys.*, *23*: 896 (1955).

55. M. L. Rogers and R. L. White, *Appl. Spectrosc.*, *41*: 1052 (1987).

56. L. N. Drozdov-Tichomirov, *Opt. Spectrosc.*, *27*: 77 (1968).

57. B. R. Kowalski, P. C. Jurs, T. L. Isenhour, and C. N. Reilley, *Anal. Chem.*, *41*: 1945 (1969).

58. R. W. Liddell III and P. C. Jurs, *Appl. Spectrosc.*, *27*: 371 (1973).

59. D. R. Preuss and P. C. Jurs, *Anal. Chem.*, *46*: 520 (1974).

60. R. W. Liddell III and P. C. Jurs, *Anal. Chem.*, *46*: 2126 (1974).

61. H. B. Woodruff, S. R. Lowry, and T. L. Isenhour, *Anal. Chem.*, *46*: 2150 (1974).

62. H. B. Woodruff, S. R. Lowry, and T. L. Isenhour, *Appl. Spectrosc.*, *29*. 226 (1975).

63. S. R. Lowry, H. B. Woodruff, G. L. Ritter, and T. L. Isenhour, *Anal. Chem.*, *47*: 1126 (1975).

64. H. B. Woodruff, S. R. Lowry, G. L. Ritter, and T. L. Isenhour, *Anal. Chem.*, *47*: 2027 (1975).

65. S. R. Lowry and T. L. Isenhour, *J. Chem. Inf. Comput. Sci.*, *15*: 212 (1975).

66. H. B. Woodruff, G. L. Ritter, S. R. Lowry, and T. L. Isenhour, *Appl. Spectrosc.*, *30*: 213 (1976).

67. G. L. Ritter, S. R. Lowry, H. B. Woodruff, and T. L. Isenhour, *Anal. Chem.*, *48*: 1027 (1976).

68. L. Domkos, I. Frank, G. Matolcsy, and G. Jalsovszky, *Anal. Chim. Acta*, *154*: 181 (1983).

69. D. S. Frankel, *Anal. Chem.*, *56*: 1011 (1984).

70. J. M. Comerford, P. G. Anderson, W. H. Snyder, and H. S. Kimmel, *Spectrochim. Acta*, *33A*: 651 (1977).

71. R. E. Dessy, *Anal. Chem.*, *56*: 1200A (1984).

72. H. B. Woodruff and M. E. Munk, *Anal. Chim. Acta*, *95*: 13 (1977).

73. H. B. Woodruff and M. E. Munk, *J. Org. Chem.*, *42*: 1761 (1977).

74. H. B. Woodruff and G. M. Smith, *Anal. Chem.*, *52*: 2321 (1980).

75. S. A. Tomellini, D. D. Saperstein, J. M. Stevenson, G. M. Smith, H. B. Woodruff, and P. F. Seelig, *Anal. Chem.*, *53*: 2367 (1981).

76. S. A. Tomellini, J. M. Stevenson, and H. B. Woodruff, *Anal. Chem.*, *56*: 67 (1984).

77. S. A. Tomellini, J. M. Stevenson, and H. B. Woodruff, *Anal. Chim. Acta.*, *162*: 227 (1984).

78. M. A. Puskar, S. P. Levine, and S. R. Lowry, *Anal. Chem.*, *58*: 1156 (1986).

79. M. A. Puskar, S. P. Levine, and S. R. Lowry, *Anal. Chem.*, *58*: 1981 (1986).

80. L. S. Ying, S. P. Levine, S. A. Tomellini, and S. R. Lowry, *Anal. Chem.*, *59*: 2197 (1987).

81. M. O. Trulson and M. E. Munk, *Anal. Chem.*, *55*: 2137 (1983).

82. D. D. Saperstein, *Appl. Spectrosc.*, *40*: 344 (1986).

83. S. Sasaki, I. Fujiwara, and T. Yamasaki, *Anal. Chim. Acta*, *122*: 87 (1980).

84. I. Fujiwara, T. Okuyama, T. Yamasaki, H. Abe, and S. Sasaki, *Anal. Chim. Acta*, *133*: 527 (1981).

85. S. Moldoveanu and C. A. Rapson, *Anal. Chem.*, *59*: 1207 (1987).

6

Applications of Chromatography/
Fourier Transform Infrared Spectroscopy

I. INTRODUCTION

Preceding chapters describe properties of chromatography/
FT-IR interfaces and outline methods for obtaining structure-
specific information from infrared spectra. This chapter contains descriptions of some mixture analysis problems to which
chromatography/FT-IR has been applied. Applications described
here are by no means the only ones for which chromatography/
FT-IR is useful. These applications are presented in order to
illustrate the advantages and limitations of chromatography/
FT-IR analysis. Sample composition, sensitivity requirements,
and analysis time should be taken into account when designing
a particular analysis method. The reader is cautioned that
unequivocal structure elucidation is generally not achieved by
chromatography/FT-IR analysis alone. Infrared analysis is
particularly useful for isomer discrimination, but is often unable
to distinguish homologues. However, additional information
provided by other detection methods and *a priori* mixture-
composition information can be combined with infrared spectroscopic analysis to obtain more accurate structure assignments.

II. TOXINS AND CARCINOGENS

Rapid worldwide industrial growth has led to generation and
disposal of large quantities of toxic substances. In an attempt
to reduce environmental pollution, legislation was passed in the
United States to regulate pollutants expelled by industry. The
resource conservation and recovery act of 1976 requires that
the U.S. Environmental Protection Agency (EPA) monitor
hazardous waste activities. In order to enforce legislation
restricting pollution, methods for analyzing hazardous waste
had to be developed. The most common analytical method in
use today for the identification of volatile and semivolatile
toxic substances is GC/MS [1]. However, GC/FT-IR has also
been employed for organic compound identification, and the
combination of complementary infrared and mass spectrometry
(GC/FT-IR/MS) shows promise for becoming a powerful and
versatile tool for rapid complex-mixture screening [2].

In 1976 the EPA established a list of "priority" pollutants
that were deemed to be toxic to living organisms and should
be regulated in wastewater effluent streams. This list contains 106 organic compounds, 9 organic mixtures (PCBs), 12
metals, cyanide ion, and asbestos. These substances were
selected based on known human and animal toxicity and carcinogenic properties. A list of the 115 organic priority pollutants is contained in Table 6-1. The pollutants are categorized by the method used for isolating them from environmental samples. Category I contains volatile organic species
that can be isolated by inert gas purge and trap procedures.
Category II species can be extracted by methylene chloride
from samples adjusted to pH 11. Category III compounds can
be extracted by methylene chloride from samples adjusted to
pH 2. Category IV materials can be extracted at ambient pH
with a mixture of methylene chloride (15%) and hexane. Gas
chromatographic analysis can be employed to monitor the 115
organic priority pollutants after they have been extracted into
a suitable solvent.

Standardized chromatographic analysis procedures were
developed by the EPA to test for the presence of priority pollutants in environmental samples. Priority pollutant screening
is most often achieved by target compound analytical methods
[3]. In target analysis, component identification is based on
comparisons with known (target) standards. Target compound
identification procedures consist of chromatographic separation
with mass spectrometry or infrared spectroscopy detection.

Comparing reference spectra with environmental sample spectra obtained by GC/MS or GC/FT-IR permits tentative compound identification. Unequivocal identification requires additional correlation between standards and unknowns. Agreement between standard and unknown chromatographic retention indices is often used to unambiguously identify target compounds. The primary drawback of the target analysis method is that only components for which standards are available can be identified and quantified unequivocally. The majority of mixture components in environmental samples are therefore ignored. Furthermore, environmental samples are too complicated to be completely resolved by capillary gas chromatography [4]. This means that spectral information obtained for some chromatographic eluents may represent mixtures of components. Complementary information derived from multiple structure-specific analysis methods may help to circumvent this problem. In fact, integrated GC/FT-IR/MS systems have shown promise for simplifying and automating nontarget environmental analysis [5,6].

Environmental mixtures often contain organic species that have structures similar to target pollutants but are not appreciably toxic. To permit proper assessment of an environmental hazard, an analysis method must be capable of differentiating between these substances and priority pollutants. Discrimination is often obtained from differences in chromatographic retention times. However, structurally similar substances (isomers) may interact with chromatographic stationary phase in virtually the same manner and elute from the chromatographic system at the same time. If this occurs, the chromatographic detector must be capable of distinguishing chromatographically unresolved substances. Fortunately, isomer discrimination is one of the strengths of infrared analysis. For example, Figure 6-1 contains spectra of the three priority pollutant dichlorobenzenes (see Table 6-1). The fingerprint region in infrared spectra of these isomers is substantially different even though the structures of the di-substituted benzenes differ only in the positions of chlorine atoms on the aromatic ring.

Determining the sensitivity of an analysis technique is the first step in drafting a new analytical procedure. Gurka and coworkers [7–10] and Shafer et al. [11] evaluated the sensitivity of GC/FT-IR for hazardous waste analysis. Table 6-2 contains light pipe GC/FT-IR identification limits for some hazardous substances commonly found in environmental samples. These identification limits were established for pure substances

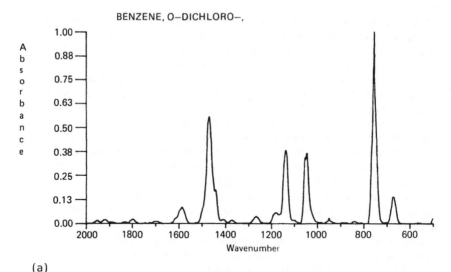

(a)

Figure 6-1   GC/FT-IR infrared spectra of (a) 1,2-dichloro-
benzene, (b) 1,3-dichlorobenzene, and (c) 1,4-dichlorobenzene.

that were analyzed by using typical environmental sample
workup procedures.   Lower identification limits were obtained
by capillary gas chromatography than by packed column gas
chromatography.   GC/FT-IR identification limits were approximately three times higher than corresponding GC/MS identification limits for strong infrared absorbers and 30 times higher
for weak absorbers [9].   Based on the applicability of GC/
FT-IR for sensitive detection and structural analysis of environmental mixture components, analysis protocol [7] and guidelines
for measuring on-the-fly GC/FT-IR reference spectra have been
established [12].

Political decisions based on environmental analyses derived
from nonrepresentative samples can have far-reaching negative
effects.   Therefore, the method of obtaining representative
environmental samples must be considered a crucial component
of an environmental assay.   Wastewater effluent streams must
be sampled at points where the level of environmental hazard
is accurately represented by pollutant concentration.   Atmos-
pheric sampling often involves some form of sample trapping or
concentration because pollutants are often present at levels

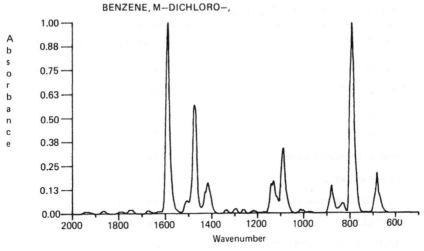

(b)

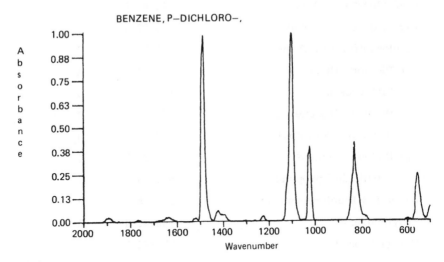

(c)

Figure 6-1 (Continued)

Table 6-1 EPA Priority Pollutants

Pollutant	CAS number[a]	Category[b]
Acrolein	107-02-8	I
Acrylonitrile	107-13-1	I
Benzene	71-43-2	I
Bromoform	75-25-2	I
Carbon tetrachloride	56-23-5	I
Chlorobenzene	108-90-7	I
Chlorodibromomethane	124-48-1	I
Chloroethane	75-00-3	I
2-Chloroethyl vinyl ether	110-75-8	I
Chloroform	67-66-3	I
Bis(chloromethyl) ether	542-88-1	I
Dichlorobromomethane	75-27-4	I
Dichlorodifluoromethane	75-71-8	I
1,1-Dichloroethane	75-34-3	I
1,2-Dichloroethane	107-06-2	I
1,1-Dichloroethylene	75-35-4	I
1,2-*trans*-Dichloroethylene	540-59-0	I
1,2-Dichloropropane	78-87-5	I
cis-1,3-Dichloropropene	542-75-6	I
trans-1,3-Dichloropropene	542-75-6	I
Ethylbenzene	100-41-4	I
Methyl bromide	74-83-9	I
Methyl chloride	74-87-3	I
Methylene chloride	75-09-2	I
1,1,2,2-Tetrachloroethane	79-34-5	I
1,1,2,2-Tetrachloroethylene	127-18-4	I
Toluene	108-88-3	I

Table 6-1 (Continued)

Pollutant	CAS number[a]	Category[b]
1,1,1-Trichloroethane	71-55-6	I
1,1,2-Trichloroethane	79-00-5	I
Trichloroethylene	79-01-6	I
Trichlorofluoromethane	75-69-4	I
Vinyl chloride	75-01-4	I
Acenaphthene	83-32-9	II
Acenaphthylene	208-96-8	II
Anthracene	120-12-7	II
Benzidine	98-87-5	II
Benzo[a]anthracene	56-55-3	II
Benzo[b]fluoranthene	205-99-2	II
Benzo[k]fluoranthene	207-08-9	II
Benzo[ghi]perylene	191-24-2	II
Benzo[a]pyrene	50-32-8	II
4-Bromophenyl phenyl ether	101-55-3	II
Butyl benzyl phthalate	85-68-7	II
Bis(2-chloroethoxy) methane	111-91-1	II
Bis(2-chloroethyl) ether	111-44-4	II
Bis(2-chloroisopropyl) ether	39,638-32-9	II
2-Chloronaphthalene	91-58-7	II
4-Chlorophenyl phenyl ether	7005-72-3	II
Chrysene	218-01-9	II
Dibenzo[a,h]anthracene	53-70-3	II
Di-n-butyl phthalate	84-74-2	II
1,2-Dichlorobenzene	95-50-1	II
1,3-Dichlorobenzene	541-73-1	II
1,4-Dichlorobenzene	106-46-7	II

Table 6-1   (Continued)

Pollutant	CAS number[a]	Category[b]
3, 3'-Dichlorobenzidine	91-94-1	II
Diethyl phthalate	84-66-2	II
Dimethyl phthalate	131-11-3	II
2, 4-Dinitrotoluene	121-14-2	II
2, 6-Dinitrotoluene	606-20-2	II
Di-*n*-octyl phthalate	117-84-0	II
1, 2-Diphenylhydrazine	122-66-7	II
Bis(2-ethylhexyl) phthalate	117-81-7	II
Fluoranthene	106-44-0	II
Fluorene	86-73-7	II
Hexachlorobenzene	118-74-1	II
Hexachlorobutadiene	87-68-3	II
Hexachlorocyclopentadiene	77-47-4	II
Hexachloroethane	67-72-1	II
Indeno[1,2,3-*cd*]pyrene	193-39-5	II
Isophorone	78-59-1	II
Naphthalene	91-10-3	II
Nitrobenzene	98-59-1	II
N-nitrosodimethylamine	62-75-9	II
N-nitrosodi-*n*-propylamine	621-64-7	II
N-nitrosodiphenylamine	86-30-6	II
Phenanthrene	85-01-8	II
Pyrene	129-00-0	II
1, 2, 4-Trichlorobenzene	129-82-1	II
p-Chloro-*m*-cresol	59-50-7	III
2-Chlorophenol	95-57-8	III
2, 4-Dichlorophenol	120-83-2	III

Table 6-1 (Continued)

Pollutant	CAS number[a]	Category[b]
2,4-Dimethylphenol	105-67-9	III
4,6-Dinitro-*o*-cresol	534-52-1	III
2,4-Dinitrophenol	51-28-5	III
2-Nitrophenol	88-75-5	III
4-Nitrophenol	100-02-7	III
Pentachlorophenol	87-86-5	III
Phenol	108-95-2	III
2,4,6-Trichlorophenol	88-06-2	III
Aldrin	309-00-2	IV
Aroclor 1016 (PCB)	12,674-11-2	IV
Aroclor 1221 (PCB)	11,104-28-2	IV
Aroclor 1232 (PCB)	11,141-16-5	IV
Aroclor 1242 (PCB)	53,469-21-9	IV
Aroclor 1248 (PCB)	12,672-29-6	IV
Aroclor 1254 (PCB)	11,097-69-1	IV
Aroclor 1260 (PCB)	11,096-82-5	IV
α-Benzenehexachloride	319-84-6	IV
β-Benzenehexachloride	319-85-7	IV
δ-Benzenehexachloride	319-86-8	IV
γ-Benzenehexachloride	58-89-9	IV
Chlordane	57-74-9	IV
4,4'-DDD	72-54-8	IV
4,4'-DDE	72-55-9	IV
4,4'-DDT	50-29-3	IV
Dieldrin	60-57-1	IV
δ-Endosulfan	959-98-8	IV
β-Endosulfan	33,213-65-9	IV

Table 6-1   (Continued)

Pollutant	CAS number[a]	Category[b]
Endosulfan sulfate	1031-07-8	IV
Endrin	72-20-8	IV
Endrin aldehyde	—	IV
Heptachlor	76-44-8	IV
Heptachlor epoxide	1024-57-3	IV
2,3,7,8-TCDD	1746-01-6	IV
Toxaphene	8001-35-2	IV

[a]Chemical abstracts number.
[b]I, purgeables; II, base/neutral extractables; III, acid extractables; IV, pesticides and polychlorinated biphenyls.

below GC/FT-IR detection limits. A great deal of effort has gone into developing appropriate environmental sampling procedures [13]. EPA guidelines for liquid and solids analysis are summarized in Figure 6-2. Relatively large samples (1 L for liquids and 50 g for solids) are obtained for analysis. Organic constituents are extracted with methylene solvent and then concentrated to 1 ml prior to chromatographic analysis. By using this procedure, estimated GC/FT-IR detection limits for wastewater pollutants range from 10 to 60 ppb and detection limits for pollutants in solids range from 200 ppb to 1 ppm. Detection limit ranges were derived from routinely attainable sensitivities of currently available GC/FT-IR light pipe interfaces and represent minimum detectable quantities [7].

GC/MI/FT-IR can provide more structural information and lower detection limits than conventional light pipe GC/FT-IR. Therefore, GC/MI/FT-IR is very useful for detection of low concentration toxins. For example, GC/MI/FT-IR has been shown to be both selective and quantitative for analysis of ethyl carbamate in beverages and food [14]. Ethyl carbamate is a product of fermentation and must be monitored because of its carcinogenic properties [15]. Ethyl carbamate can be quantified in beverages by using infrared analysis with isotopically labeled (^{13}C, ^{15}N) ethyl carbamate as an internal standard.

Table 6-2 GC /FT-IR Identification Limits for Hazardous Organics

Compound	Identification limit[a]	
	ng Injected	μg /L
Isophorone	40	20
Nitrobenzene	25	12.5
Dimethyl phthalate	20	10
Dibenzofuran	40	20
2,4-Dinitrotoluene	20	10
N-nitrosodimethylamine	20	10
1,3-Dichlorobenzene	50	25
Diethyl phthalate	20	10
4-Chlorophenyl phenyl ether	20	10
Di-n-butyl phthalate	20	10
Di-n-propyl phthalate	25	12.5
Butyl benzyl phthalate	25	12.5
2-Methylnaphthalene	110	55
1,4-Dichlorobenzene	50	25
Bis(2-chloroethyl) ether	70	35
Hexachloroethane	50	25
4-Chloroaniline	40	20
2-Nitroaniline	40	20
3-Nitroaniline	40	20
4-Nitroaniline	40	20
1,2,4-Trichlorobenzene	50	25
Naphthalene	40	20
2-Chloronaphthalene	110	55
2,6-Dinitrotoluene	20	10
Bis(2-chloroisopropyl) ether	50	25

Table 6-2   (Continued)

| Compound | Identification limit[a] | |
	ng Injected	μg/L
Bis(2-chloroethoxy)methane	50	25
4-Bromophenyl phenyl ether	40	20
N-nitrosodipropylamine	50	25
N-nitrosodiphenylamine	40	20
1,2-Dichlorobenzene	50	25
Acenaphthene	40	20
Acenaphthylene	50	25
1,3-Hexachlorobutadiene	120	60
Fluorene	40	20
Anthracene	40	20
Hexachlorobenzene	40	20
Hexachlorocyclopentadiene	120	60
Phenanthrene	50	25
Fluoranthene	100	50
Pyrene	100	50
Phenol	70	35
2-Chlorophenol	50	25
2-Cresol	50	25
4-Cresol	50	25
2-Nitrophenol	40	20
Benzoic acid	70	35
2,4-Dichlorophenol	50	25
4-Chlorophenol	100	50
2,4,6-Trichlorophenol	120	60
2,4,5-Trichlorophenol	120	60
2,4-Dinitrophenol	60	30
4,6-Dinitro-2-cresol	60	30

[a]Based on a 2 μL injection of a 1-L sample that had been extracted and concentrated to a volume of 1.0 mL.
Source: Reprinted with permission from Ref. 9.   Copyright ©
1987 American Chemical Society, Washington, D.C.

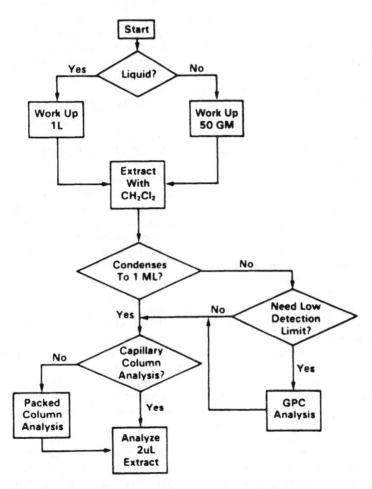

Figure 6-2   Protocol for environmental sample workup and analysis. (Reprinted with permission from Ref. 7. Copyright © 1985 Society for Applied Spectroscopy, Frederick, Maryland.)

The analysis procedure is illustrated in Figure 6-3. Figure 6.3(a) is a chromatogram of a whiskey extract with the ethyl carbamate elution marked. The retention time for labeled internal standard was identical to unlabeled ethyl carbamate. As a result, the GC/MI/FT-IR ethyl carbamate spectrum obtained from the chromatogram in Figure 6-3(a) contained infrared absorption features for both species. Asymmetric $C-O-CO$ stretching vibrations for labeled (1298 cm^{-1}) and unlabeled (1326 cm^{-1}) ethyl carbamate were well resolved in matrix-isolation spectra even though they were separated by only 28 cm^{-1} [Figure 6-3(b)]. Spectroscopic resolution of ethyl carbamate and internal standard facilitated a reliable quantitative assay even though analyte and internal standard were chromatographically unresolved. Because the internal standard for this analysis had chemical properties that were identical to the analyte, there was no danger that quantitative results would be skewed by differences in recovery during sample workup.

Results of whiskey extract analyses obtained by GC/MI/FT-IR and GC/MS/MS are compared in Table 6-3. GC/MI/FT-IR and GC/MS/MS results agree quite well, demonstrating that GC/MI/FT-IR provides adequate quantitative as well as excellent qualitative analysis capabilities for these samples.

Tetrachlorodibenzo-*p*-dioxins (TCDDs) have received considerable attention because of the extreme toxicity of some isomers. The structure of dibenzo-*p*-dioxin is shown in Figure 6-4 along with the substituent numbering system used to designate chlorinated isomers. TCDDs are stable compounds with varying mutagenic and carcinogenic properties. TCDD isomer differentiation is desired because these isomers differ significantly in toxicity. For instance, the 2,3,7,8-TCDD isomer is approximately one million times more toxic than the 1,2,3,4-TCDD isomer. TCDDs have been found in herbicides, chlorinated phenols, and other polychlorinated substances. In addition, they may be formed by combustion processes and can be found in fly ash and flue gases from municipal incinerators [16]. The 22 TCDD isomers cannot be separated by using a single gas chromatographic column [17] and isomers cannot be differentiated by low-resolution mass spectrometry [18]. In spite of this, GC/MS has been the principal method employed for analysis of TCDDs in environmental samples [19]. Alternately, GC/MI/FT-IR can be used to identify separated polychlorinated dibenzo-*p*-dioxin isomers [20−24]. All 22 TCDD isomers have been studied by GC/MI/FT-IR [21]. Each isomer is readily identified from its matrix-isolation spectra. For

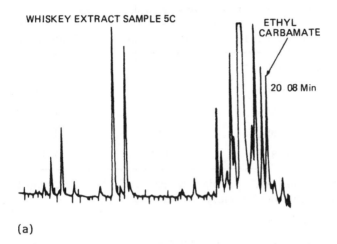

(a)

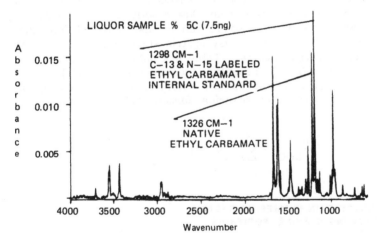

(b)

Figure 6-3 (a) FID chromatogram for a whiskey extract. (b) FT-IR spectrum of ethyl carbamate obtained by GC/MI/FT-IR. (Reprinted with permission from Ref. 14. Copyright © 1988 American Chemical Society, Washington, D.C.)

Table 6-3   Ethyl Carbamate Levels in Whiskey (ppb)

Sample	GC /MI /FT-IR	GC /MS /MS	Ratio IR /MS
2C	117	116	1.00
3C	126	108	1.17
4C	68	60	1.13
5C	389	330	1.18
6C	183	166	1.10

Source: Reprinted with permission from Ref. 14. Copyright ©
1988 American Chemical Society, Washington, D.C.

example, Figure 6-5 compares GC /MI /FT-IR spectra for the
2,3,7,8-TCDD and 1,2,3,4-TCDD isomers. These spectra contain significant differences and are readily distinguished.

Most of the TCDD isomers can be distinguished from the
position of the C—O—C asymmetric stretching band in matrix-
isolation infrared spectra. Table 6-4 lists estimated and
measured C—O—C absorption frequencies obtained by GC /
FT-IR for the 22 TCDD isomers. Estimated vibration frequencies
were obtained from an empirical formula [22]

$$v_{COC} = v^0_{COC} - \sum_{i=1}^{4} n_i S_i - F + Q \qquad (6.1)$$

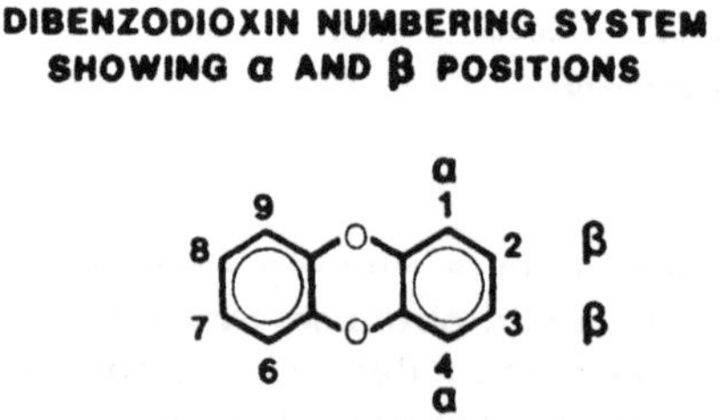

**DIBENZODIOXIN NUMBERING SYSTEM
SHOWING α AND β POSITIONS**

Figure 6-4   Structure of dibenzo-*p*-dioxin. (Reprinted with
permission from Ref. 22. Copyright © 1988 Society for Applied
Spectroscopy, Frederick, Maryland.)

Table 6-4 Estimated and Observed C—O—C Asymmetric Stretching Frequencies

TCDD isomer	GC /MI /FT-IR	GC /FT-IR	GC /FT-IR (estimated)
2378	1330	1321	1311[a]
	1313	1306	
1237	1316	1308	1308
1238	1315	1306	1306
1278	1311	1305	1305
1236	1312	1304	1303
1239	1311	1303	1301
1378	1309	1302	1302
1267	1311	1301	1301
1289	1307	1298	1298
1268	1308	1298	1300
1279	1306	1296	1296
1478	1300	1295	1293
1279	1304	1293	1293
1368	1304	1292	1292
1269	1298	1289	1290
1234	1298	1288	1288
1369	1293	1286	1285
1246	1295	1286	1286
1249	1293	1285	1285
1248	1293	1286	1287
1247	1294	1284	1286
1469	1287	1281	1282

[a]Average of overlapping bands.
Source: Reprinted with permission from Ref. 22. Copyright © 1988 Society for Applied Spectroscopy, Frederick, Maryland.

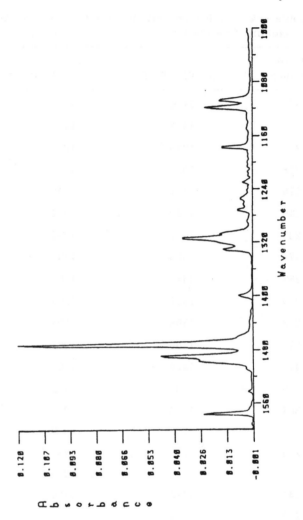

(a)

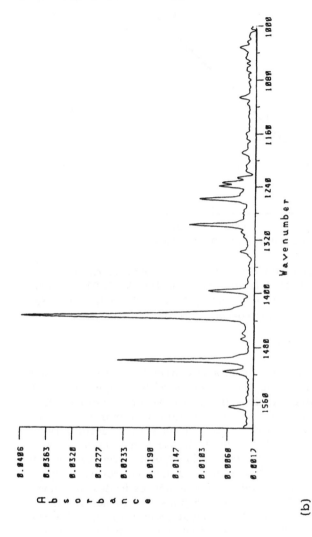

(b)

Figure 6-5 (a) GC/MI/FT-IR infrared absorbance spectrum of 2,3,7,8-tetra-chlorodibenzo-*p*-dioxin. (b) GC/MI/FT-IR infrared absorbance spectrum of 1,2,3,4-tetrachlorodibenzo-*p*-dioxin.

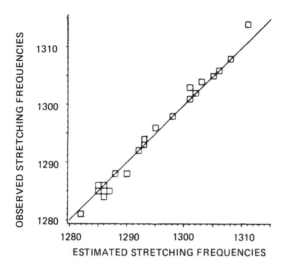

Figure 6-6   Plot of observed C—O—C asymmetric stretching frequency vs estimated frequencies for the 22 TCDD isomers. (Reprinted with permission from Ref. 22.   Copyright © 1988 Society for Applied Spectroscopy, Frederick, Maryland.

In Eq. (6.1) v^0_{COC} represents the stretching frequency for the molecule with the highest electron-withdrawing ring.   The number of possible oxonium ion delocalization structures of type i is denoted by n_i. S_i is a weighting factor for oxonium ion delocalization structures that varies with structural type. Four types are defined for which S_i has values of 0, 1, 2, and 3 cm^{-1}.   F and Q are correction factors used to compensate for 1,9 and 1,4 substituent interactions, respectively. Empirical C—O—C stretching frequency estimates are very close to observed GC/FT-IR vapor phase frequencies.   In fact, a plot of observed vs estimated band frequency for the 22 isomers (Figure 6-6) has a correlation coefficient of 0.99 [22].

Figure 6-7 illustrates the use of GC/MI/FT-IR for analysis of a soil extract previously analyzed by GC/MS and found to contain about 40 ppm 2,3,7,8-TCDD.   As if often the case, gas chromatography could not separate the 2,3,7,8-TCDD isomer from co-eluting substances.   By subtracting GC/FT-IR spectra corresponding to leading [Figure 6-7(a)] and trailing

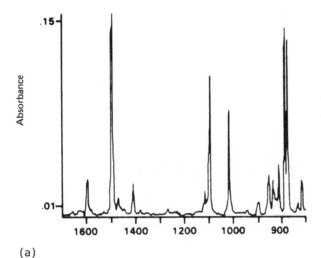

(a)

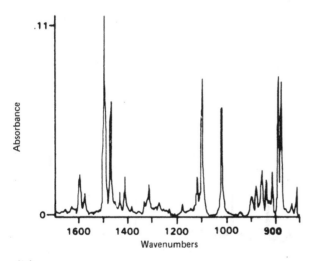

(b)

Figure 6-7   GC /MI /FT-IR infrared absorbance spectra obtained from a soil extract analysis and measured at the
2,3,7,8-TCDD retention time.   (a) Spectrum obtained from
leading edge of GC peak.   (b) Spectrum obtained from trailing edge of GC peak.   (c) Difference spectrum calculated by
subtracting (b) from (a).   (d) Reference GC /MI /FT-IR spectrum of 2,3,7,8-TCDD.   (Reprinted with permission from
Ref. 21.   Copyright © 1986 Society for Applied Spectroscopy,
Frederick, Maryland.)

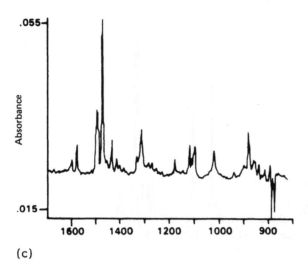

(c)

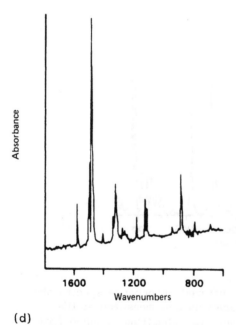

(d)

Figure 6-7   (Continued)

[Figure 6-7(b)] edges of the broad chromatographic peak, an identifiable 2,3,7,8-TCDD spectrum was obtained [Figure 6-7(c)]. Figure 6-7(d) is a reference spectrum of 2,3,7,8-TCDD for comparison.

III. WASTEWATER CONSTITUENTS

Shafer and coworkers employed complementary mass spectrometric and infrared detection with gas and liquid chromatography to analyze wastewater samples [25,26]. One of the wastewater samples analyzed in their studies was obtained from a chemical plant that manufactured nitrobenzene, dichlorobenzene, o-nitrophenol, aniline, and various oil additives [25]. They identified nine substituted aromatic components in the wastewater extract as p-dichlorobenzene, m-chloronitrobenzene, p-chloronitrobenzene, o-chloronitrobenzene, 3,4-dichloronitrobenzene, 2,4-dichloro-6-nitroaniline, trichloronitrobenzene, alkyl phthalate, and triphenylphosphate. GC/MS was unable to discriminate between isomeric chloronitrobenzenes, These assignments were made by complementary information provided by GC/FT-IR analysis. GC/FT-IR analysis identified one component as either trichloronitrobenzene or 2,3-dichloronitroaniline. Additional information provided by GC/MS led to identification of this component as being trichloronitrobenzene.

Azarraga and Potter developed a novel GC/FT-IR analysis system for nondestructive environmental analysis [27]. A diagram of this system in shown in Figure 6-8. Nondestructive FT-IR detection permits eluent trapping and concentrating at various stages of separation. Sample introduction was made by direct injection or thermal desorption from adsorbent cartridges. The apparatus incorporated two different gas chromatographic columns for mixture-component separations. One of these columns was an SP-2250 SCOT column and the other was an OV-101 WCOT column. Figures 6-9 and 6-10 are GC/FT-IR chromatograms for ink- and printing-industry wastewater effluent extracts separated by these columns. Neither column resolved all of the extract components. By using the apparatus shown in Figure 6-8, components eluting after 850 sec on the SP-2250 column were isolated and trapped. Trapping was repeated four times to concentrate these substances. The trapped components were then flash-vaporized onto the OV-101 column, producing the chromatogram and corresponding

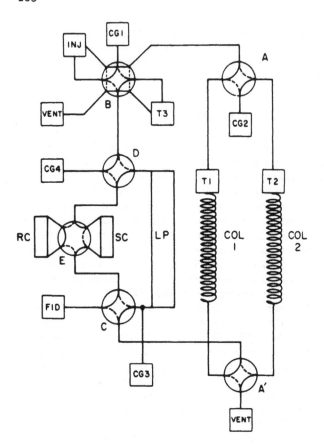

Figure 6-8   Multipurpose GC/FT-IR light pipe interface.
Valves A (A and A') and B have 8 ports, C and D 4 ports,
and E 6 ports.   Helium carrier gas is supplied by the GC
capillary source (CG1), two flow-regulated sources (CG2 and
CG3), and a regulated helium tank (CG4).   Other indicated
interface components are INJ (capillary injector port); T1, T2,
T3 (cold traps); COL1 and COL2 (capillary columns); FID
(flame ionization detector); SC (sample cartridge); RC
(recovery cartridge); and the LIGHT PIPE.   (Reprinted from
Ref. 27.   Copyright © 1981 Alfred Heuthig Publications, New York.)

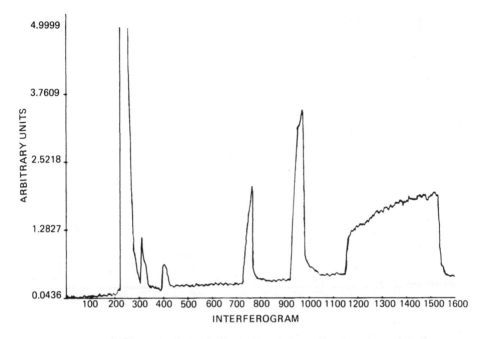

Figure 6-9   GC /FT-IR chromatogram for an extract prepared from ink and printing industrial wastewater effluent.   An SP-2250 column was used for separation of components. (Reprinted from Ref. 27.   Copyright © 1981 Alfred Heuthig Publications, New York.)

infrared spectra shown in Figure 6-11.   Detection sensitivity and chromatographic resolution were enhanced by using the dual column separation method (compare the chromatogram in Figure 6-11 with Figure 6-10).   The first column of infrared spectra in Figure 6-11 corresponds to components eluting earlier than 850 sec on the OV-101 column.   Spectrum 263 was identified as water vapor that was probably trapped with the sample during cooling.   Spectra for files 412 and 446 match aliphatic ketone library spectra.   For example, the best search match for spectrum 412 was 5-methyl-2-hexanone.   Spectrum 522 was identified as a siloxane that was a known contaminant in the laboratory air and presumably trapped accidentally. The second column of spectra represents components eluting

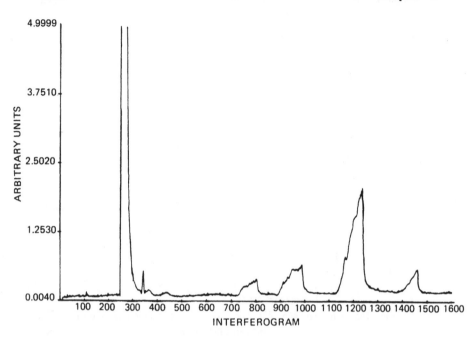

Figure 6-10   GC/FT-IR chromatogram for an extract prepared
from ink and printing industrial wastewater effluent.   An
OV-101 column was used to separate components.   (Reprinted
from Ref. 27.   Copyright © 1981 Alfred Heuthig Publications,
New York.)

between 900 and 1000 sec on the OV-101 column.   These spectra are indicative of carbitol ($C_2H_5OCH_2CH_2OH$) derivatives.
The third and fourth columns of spectra characterize components
eluting between 1080 and 1200 sec on the OV-101 column and
indicate butoxyethoxyethanols.   From this data, Azarraga and
Potter concluded that butoxyethoxyethanol derivatives were
the major constituents of the wastewater extract.

Malissa et al. employed capillary GC/FT-IR for the analysis
of chlorinated phenols in surface waters [28].   Surface-water
chlorophenols arise from herbicide and insecticide degradation,
drinking water chlorination, and industrial effluent reactions.
Infrared spectroscopy was selected for analysis because it can
distinguish isomeric chlorinated phenols more readily than mass
spectroscopy.   Correct identifications were made for more than

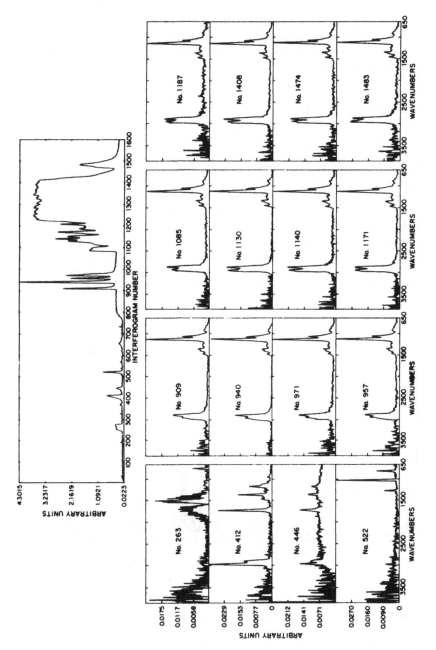

Figure 6-11 GC/FT-IR chromatogram and representative vapor phase infrared spectra for trapped material isolated from the SP-2250 column and separated on the OV-101 column. (Reprinted from Ref. 27. Copyright © 1981 Alfred Heuthig Publications, New York.)

20 chlorophenols at 10 ppb levels after samples were concentrated by a factor of 10^4. Spiked sample analysis indicated that mean recoveries of about 90% could be obtained for the sample preparation method employed.

IV. SEDIMENT EXTRACTS

Gurka and coworkers [6,29—31] and Shafer et al. [32] employed GC/FT-IR, GC/MS, and integrated GC/FT-IR/MS for analysis of soil and chemical still-bottom extracts. Gurka et al. analyzed soil samples from an abandoned chemical dump site and an area contaminated by agricultural pesticides. Acetonitrile was used to extract organics from the soil samples. Organic species were then transferred from acetonitrile to isooctane that served as the solvent for gas chromatographic analysis. GC/MS and GC/FT-IR reconstructed chromatograms for one of the soil sample extracts are shown in Figure 6-12. The chemical dump-site soil was found to contain isomeric chlorinated aromatics.

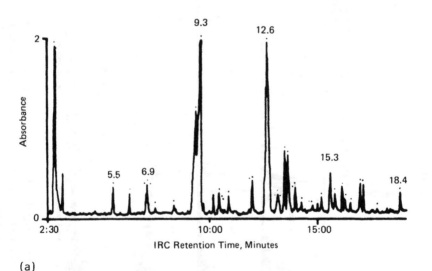

(a)

Figure 6-12   GC/FT-IR chromatogram (a) and GC/MS chromatogram (b) for a sediment extract. (Reprinted with permission from Ref. 30. Copyright © 1984 American Chemical Society, Washington, D.C.)

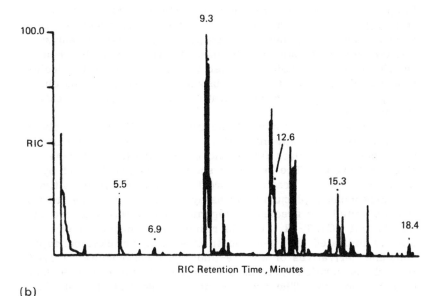

(b)

Figure 6-12 (Continued)

Spectra were obtained for a total of 43 components in the soil sample. Of these, chemical class identifications were made for 32 and unequivocal identifications were made for 23 based on interpretation of infrared and mass spectra. Nine of the components were found to be EPA priority pollutants.

In a related study, Gurka and Titus used integrated GC/ FT-IR/MS for nontarget analysis of sediment extracts [5]. Six environmental samples were analyzed, including three dyes, a pesticide-contaminated soil, a chlorohydrocarbon-contaminated soil, and a herbicide still-bottom. From all sample extracts, 106 analytes were jointly detected by mass spectrometry and infrared spectroscopy. Of these, 24 compound class and 20 unambiguous identifications were made. Confirmed structural information was reported for 41% of all detectable species.

V. AIRBORNE SPECIES

GC/FT-IR analysis of airborne organics is particularly difficult because target species are often found in low concentrations.

Sample preparation for this type of analysis usually involves trapping organics on an absorbent material from which it can later be released by heating. To permit quantitative analysis, a known volume of air must be passed through the absorbent cartridge during sampling. After trapping, GC/FT-IR injection is achieved by connecting the absorbent cartridge to the chromatographic column and rapidly heating the cartridge to release trapped species.

Erickson et al. reported the use of GC/FT-IR for analysis of diesel exhaust pollutants [33]. Samples of diesel exhaust were collected on glass fibers and extracted with methylene chloride solvent. The resulting solution was fractionated by HPLC, and various fractions were tested for mutagenic activity in a Salmonella typhimurium bioassay (Ames test). The fraction that exhibited the greatest mutagenic activity was analyzed by GC/MS and GC/FT-IR. GC/MS results were inconclusive. Spectra indicated that isomeric species could be alkyl benzo[c]-cinnoline derivatives or alkyl-9-fluorenone homologues. GC/FT-IR analysis verified that mixture components were alkyl-9-fluorenone homologues. The presence of a carbonyl stretch at 1723 cm^{-1} in GC/FT-IR spectra was evidence that components could not be alkyl benzo[c]cinnoline derivatives. There is precedence for the presence of fluorenones in diesel exhaust. For example, alkyl-9-fluorenones have also been detected in crude oil and cigarette smoke condensates [34,35].

Polycyclic aromatic hydrocarbons (PAHs) and their derivatives are often found in emissions from gasoline and diesel engines [36]. Nitro-PAHs are thought to be formed by the reaction of nitrogen oxides and nitric acid with PAH materials present in the fuels [37]. Like PAHs, nitro-PAHs have been found to possess carcinogenic properties [38,39]. Kalasinsky et al. developed a GC/FT-IR method for PAH analysis and selective detection of nitro-PAHs in diesel fuel combusion emissions [40]. A 30m × 0.25mm I.D. SE-30 capillary column was employed for separations. Symmetric and asymmetric NO_2 stretching-vibration absorptions in vapor phase infrared spectra from 1320 to 1380 cm^{-1} and 1520 to 1600 cm^{-1}, respectively, were integrated during chromatographic separation to provide selective detection of nitro-PAHs. Spectra for isomeric nitro-PAHs shown in Figure 6-13 exhibit maximum absorbance for the NO_2 symmetric and asymmetric stretches. The 3010 to 3080 cm^{-1} and 720 to 800 cm^{-1} ranges were also integrated during separation because these regions are representative of PAHs in general. A chemigram for the separation of 13 PAHs is shown in Figure 6-14.

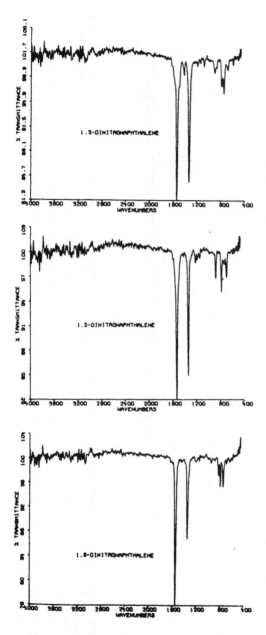

Figure 6-13 GC/FT-IR vapor phase spectra for isomeric nitro-
PAHs. (Reproduced from Ref. 40 by permission of Preston
Publications, a division of Preston Industries, Inc., Niles,
Illinois.)

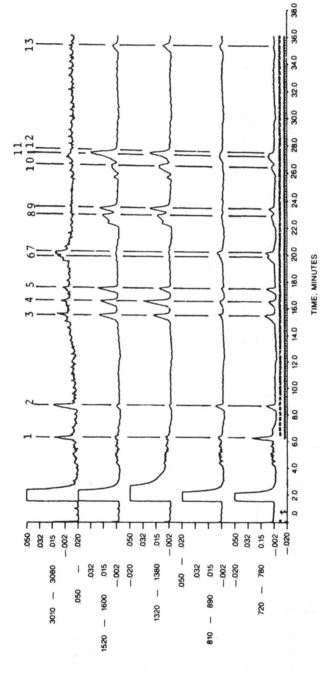

Figure 6-14 A chemigram reconstructed GC/FT-IR chromatogram for 13 PAH materials. (1) naphthalene; (2) 2-methyl-naphthalene; (3) 1-nitronaphthalene; (4) 2-nitronaphthalene; (5) 2-nitrobiphenyl; (6) phenanthrene; (7) anthracene; (8) 1,5-dinitronaphthalene; (9) 1,3-dinitronaphthalene; (10) 2-nitro-fluorene; (11) 1,8-dinitronaphthalene; (12) 9-nitroanthracene; (13) 1-nitropyrene. (Reproduced from Ref. 40 by permission of Preston Publications, a division of Preston Industries, Inc., Niles, Illinois.)

Table 6-5   Identified Compounds in Coal Gasification Sample

File no.	Compound identification		
	Manual	Search	Final
451	Methylene chloride	methylene chloride	methylene chloride
455	Benzene	benzene	benzene
465	Toluene	toluene	toluene
474	m-Xylene (tent.)	m-xylene	m-xylene
480	o-Xylene (tent.)	o-xylene	o-xylene
486	Ethylbenzene (tent.)	alkylbenzene and halomethanes	alkylbenzene (tent.)
490	1,3,5-Trimethylbenzene	alkanes	1,3,5-trimethylbenzene
495	1,2,4-Trimethylbenzene	1,2,4-trimethylbenzene	1,2,4-trimethylbenzene
502	Alkylbenzene	—	alkylbenzene
507	Phenol	phenol	phenol
521	Indan	—	indan
526	Indene	indene	indene
532	o-Cresol and p-cresol	phenolics	o-cresol
539	m-Cresol	m-cresol	m-cresol

Table 6-5 (Continued)

File no.	Compound identification		
	Manual	Search	Final
544	Alkylbenzene	—	alkylbenzene
548	2,6-Xylenol	—	alkylphenol
553	2,4-Xylenol	—	alkylphenol
560	Naphthalene	naphthalene	naphthalene
568	1-Methylnaphthalene (tent.)	benzothiofuran	benzothiofuran
589	2-Methylnaphthalene	isoquinoline	dimethylnaphthalene (tent.)
594	Alkyl aromatic	—	acid (tent.)
598	—	1-methylnaphthalene	1-methylnaphthalene
617	Alkyl aromatic	—	acid (tent.)
625	—	biphenyl	biphenyl
661	Ester	esters/alcohols	ester
708	Dimethylnaphthalene	alkyl aromatic	alkylnaphthalene
720	1,4-Dimethylnaphthalene	alkyl aromatic	1,4-dimethylnaphthalene
775	—	phenanthrene	phenanthrene

Source: Reprinted with permission from Ref. 42. Copyright © 1981 Society for Applied Spectroscopy, Frederick, Maryland.

The nitro-PAHs (3,4,5,8,9,10,11,12,13) were readily distinguished from PAHs (1,2,6,7) by the 1520 to 1600 cm^{-1} and
1320 to 1380 cm^{-1} integration windows. High nanogram nitro-
PAH detection limits were attained for 1 μl sample injections.

Volatile organic pollutants produced by a semibatch bench-
scale coal gasifier were characterized by GC/FT-IR [41,42].
Emissions from the gasifier were trapped in a XAD-2 resin and
subsequently analyzed by GC/FT-IR. Library searching was
employed to facilitate component identification. Identified components of coal gasifier emission are compiled in Table 6-5.
Manual identifications were made by visually comparing GC/FT-
IR spectra with reference spectra. Search identifications listed
in Table 6-5 were the best library match. The best match was
either the closest match or the closest logical match from the
top five search matches. Several of the identified components
were found to be EPA priority pollutants.

VI. PESTICIDES AND THEIR DEGRADATION PRODUCTS

Insecticide formulations are developed with at least three considerations in mind. First, the formulation must effectively
control the pest it was designed for. Second, the insecticide
must be easily distributed in areas populated by the pest.
Third, the insecticide must degrade rapidly in the environment
after it has completed its designed task and degradation products
should be inert. Kalasinsky has shown that GC/FT-IR can be a
useful analytical tool for developing pesticide formulations
[43 – 45]. GC/FT-IR was used to characterize degradation
products formed in the laboratory and in the field from a mirex
formulation targeted to control the fire ant pest [43,44]. Electron-capture gas chromatography and GC/MS were found to be
more sensitive than GC/FT-IR for degradation product analysis,
but these detectors provided ambiguous identifications.

Sunlight decomposes mirex (Figure 6-15) into hydrogen-
substituted derivatives. Four major products were obtained
from controlled photodecomposition of the pesticide (Figure
6-16). These components were identified from their infrared
spectra as 2,8-dihydrogen mirex, anti-5,10-dihydrogen mirex,
8-monohydrogen mirex, and 10-monohydrogen mirex (Figure
6-17). After pesticide degradation products were identified,
they were isolated and tested for toxicity. GC/FT-IR can also
be used for testing the purity of isolated products. For
example, toxicity tests of two aliquots of presumably the same
mirex derivative yielded conflicting results [45]. One of the

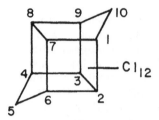

Figure 6-15   Molecular structure for mirex. (Reprinted with permission from Ref. 44.   Copyright © 1983 Preston Publications, a division of Preston Industries, Inc., Niles, Illinois.)

aliquots was found to be carcinogenic, but the other was not. Electron-capture gas chromatographic analysis of the derivatives indicated that both isolated products were pure.   However, GC/FT-IR analysis revealed that one of the samples was impure and was contaminated with triphenylphosphine oxide, which is a known carcinogen.   This apparent discrepancy resulted because electron-capture detectors are relatively insensitive to organo-phosphorous compounds.   In contrast, this compound class is readily detected and identified by GC/FT-IR.

Sometimes, pesticides or their degradation products cannot be analyzed by gas chromatography because they are nonvola-tile or thermally labile.   In these instances, LC/FT-IR tech-niques can be used for structural analysis.   For example, HPLC/FT-IR has proven useful for analysis of mixtures con-taining the pesticide Erythrosin B.   Erythrosin B can be used to control flies commonly found in chicken coups.   Young flies have translucent bodies.   After ingesting the dye, sunlight causes the dye inside the insects to photooxidize, which kills them.   Figure 6-18 shows spectra obtained during a reverse phase HPLC/DRIFTS analysis in which Erythrosin B dye was being separated [45].   Methanol was employed as the reverse phase eluant.   DFN values in Figure 6-18 denote the number of the acquired spectrum relative to the start of the separation. Identifiable Erythrosin B infrared spectra were obtained for five of the HPLC/FT-IR data files.

Holloway et al. employed GC/MI/FT-IR to study the environ-mental degradation of Fonofos and Terbufos (Figure 6-19) [23]. The goal of the study was to determine the persistence of these pesticides in the environment and find a method for distinguishing

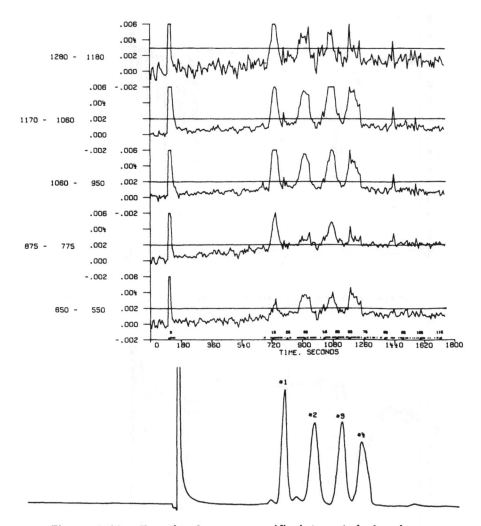

Figure 6-16 Functional group-specific integrated absorbance chromatogram (top) and thermal conductivity chromatogram (bottom) for a mirex photodecomposition product separation. (Reproduced from Ref. 44 by permission of Preston Publications, a division of Preston Industries, Inc., Niles, Illinois.)

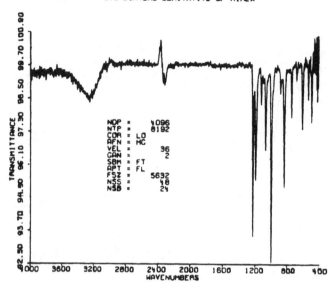

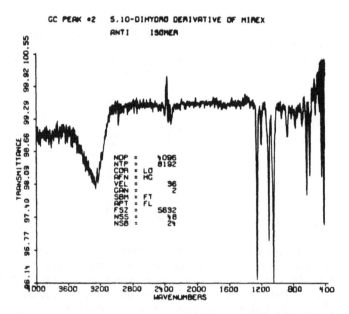

Figure 6-17   Vapor phase GC/FT-IR infrared spectra for gas chromatographic peaks numbered in Figure 6-16.   (Reproduced from Ref. 44 by permission of Preston Publications, a division of Preston Industries, Inc., Niles, Illinois.)

GC PEAK #3   8-MONOHYDRO DERIVATIVE OF MIREX

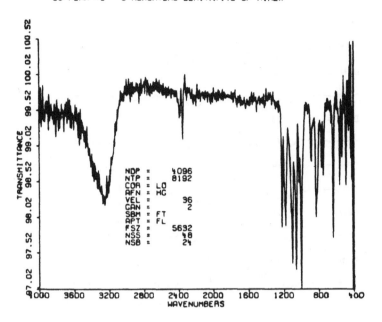

GC PEAK #4   10-MONOHYDRO DERIVATIVE OF MIREX

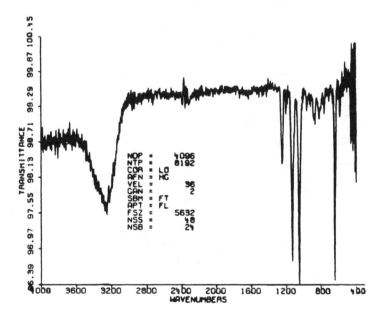

Figure 6-17   (Continued)

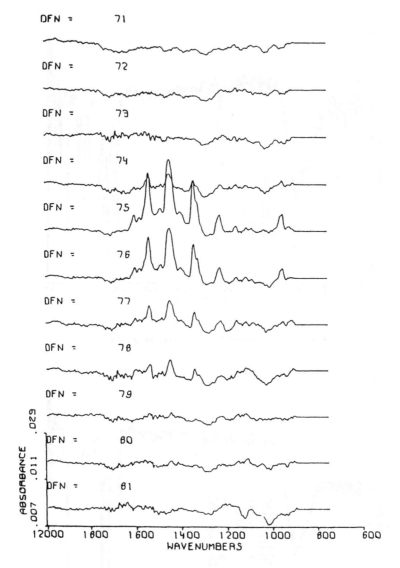

Figure 6-18   LC/FT-IR infrared spectra obtained for a separation of Erythrosin B. (Reprinted from Ref. 45. Copyright © 1985 John Wiley & Sons, New York.)

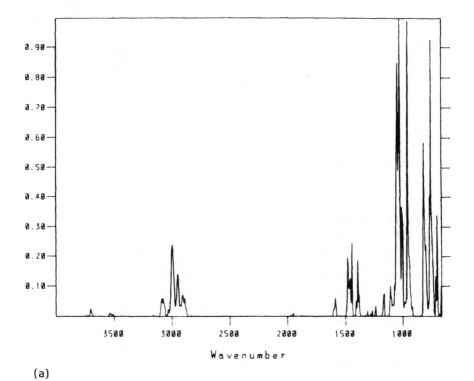

(a)

CH₃CH₂—O, S
         \ ‖
          P=S
         /
CH₃CH₂   S—⬡

(b)

CH₃CH₂—O S   CH₃
        \ ‖ |
         P—S—CH₂—S—C—CH₃
        / |
CH₃CH₂—O   CH₃

Figure 6-19   Molecular structures of (a) Fonofos and (b) Terbufos. (Reprinted with permission from Ref. 23. Copyright © 1988 Society for Applied Spectroscopy, Frederick, Maryland.)

(a)

Wavenumber

Figure 6-20   GC/MI/FT-IR spectra for (a) Fonofos and (b) Terbufos. (Reprinted with permission from Ref. 23. Copyright © 1984 Society for Applied Spectroscopy, Frederick, Maryland.)

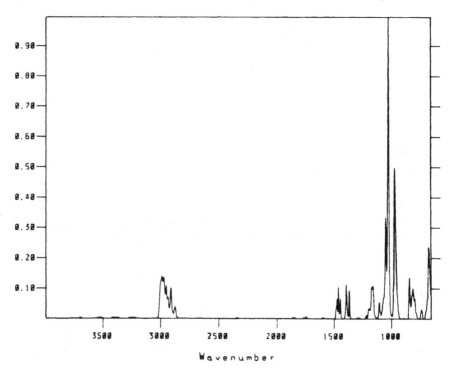

(b)

Figure 6-20   (Continued)

these substances in groundwater samples. GC /MI /FT-IR spectra
for these pesticides have significant differences and are readily
differentiated (Figure 6-20). Degradation studies showed that
Terbufos decomposed rapidly in river water, reaching undetect-
able levels in 11 days. In contrast, Fonofos concentration
decreased to one-half the initial concentration in about one
month. Apparently, the aromatic ring in the Fonofos molecule
stabilizes this species.

VII. FUELS AND FUEL FEEDSTOCKS

The 1970s energy crisis in the United States led to increased
efforts to conserve petroleum resources and search for

alternate fuel sources. Petroleum-derived fuels are extremely complex mixtures. Gasoline, for example, contains over 200 components [46]. Many of these components are isomeric species. GC/FT-IR is unsurpassed for analysis of isomeric mixtures and is well suited for the characterization of fuels. GC/FT-IR has been used to characterize petroleum distillates [47,48], gasoline [49], and jet fuel [50]. In addition, pyrolysis GC/FT-IR has been employed to study the effect of boron salts on the burning rate of nitramine gun propellants [51].

Cooper and Taylor analyzed jet fuel by capillary GC/FT-IR with on-column injection [50]. Figure 6-21 is an FID chromatogram for a jet fuel mixture separated by a 60 m × 0.33 mm I.D. DB-5 (1 μm film) fused silica column. Figure 6-22 is the corresponding GC/FT-IR-reconstructed chromatogram for the same mixture made by integrating infrared absorbances for acquired data files between 3150 and 2850 cm^{-1}. Relative chromatographic peak intensities are different in Figures 6-21 and 6-22 because the mechanism by which the FID and FT-IR detectors respond to separated components is different. A minor component in the jet fuel mixture was studied in order to estimate GC/FT-IR detection limits. The minor component elution is designated by a (+) in the chromatograms shown in Figures 6-21 and 6-22. The vapor phase infrared spectrum for this species is shown in Figure 6-23. Most of the infrared absorbance for this material is concentrated in the aliphatic C —H stretching region, which is characteristic of alkanes. By comparing the spectrum in Figure 6-23 to a reference spectrum of nonane measured with the same apparatus under the same conditions (Figure 6-24), a detection limit of about 40 ng (on-column) was estimated for the jet fuel GC/FT-IR analysis. Library search was used to determine chemical classes of substances separated in the jet fuel. Library-search results for one of the components (marked ++ in Figures 6-21 and 6-22) indicated that the substance was *m*-xylene (Table 6-6). By comparing this jet fuel infrared spectrum with *m*-xylene reference spectra, it was estimated that this species was present in the injected sample at a level of 400 ng. Unequivocal identification of the other mixture components could not be obtained from vapor phase library-search results alone.

Polycyclic aromatic hydrocarbons (PAHs) are often found in petroleum distillates. Some of these compounds are well-known carcinogens. For this reason, methods have been developed to monitor these substances in the environment [40]. PAHs are also particularly effective for blocking catalyst

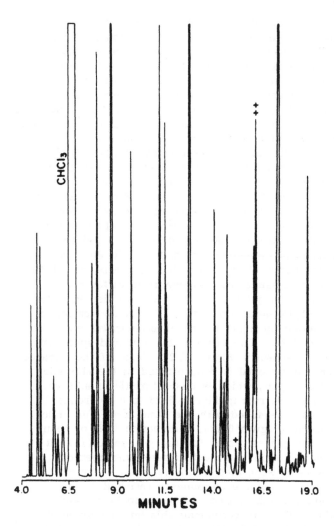

Figure 6-21   FID chromatogram for a jet fuel mixture. (Reprinted with permission from Ref. 50.   Copyright © 1984
Society for Applied Spectroscopy, Frederick, Maryland.)

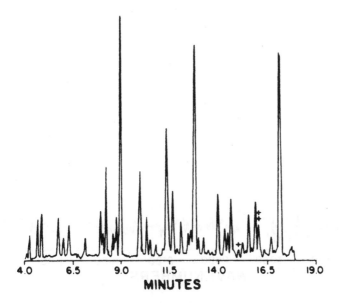

MINUTES

Figure 6-22   Functional group-specific integrated absorbance (3150−2850 cm^{-1}) GC /FT-IR chromatogram for a jet fuel sample. (Reprinted with permission from Ref. 50. Copyright © 1984 Society for Applied Spectroscopy, Frederick, Maryland.)

activity in fuel-cracking plants [52].   These materials can be converted into coke, with 85−95% yield at elevated temperatures [53].   This coke can deposit on silica- or alumina-supported catalysts and block reactive sites, rendering the catalyst less effective.   For this reason, the petroleum industry is interested in developing methods for rapid analysis of feedstocks used in catalytic cracking plants for PAH content.   Garg et al. developed a procedure in which PAHs were analyzed in petroleum feed-stock distillates by using GC /MS and GC /FT-IR [54].   GC /MS was employed for initial identification, and GC /FT-IR results were used to distinguish isomeric species that could not be differentiated by GC /MS.   GC separation with flame ionization detection was employed to quantify separated components.   By using the combined information provided by these three analysis methods, structural information and sample concentration esti-mates were made for 112 components of a petroleum distillate.

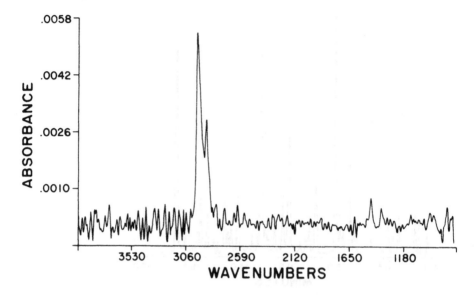

Figure 6-23   GC/FT-IR spectrum of a minor component in the jet fuel spectrum. This spectrum was obtained from chromatogram position marked (+) in Figures 6-19 and 6-20. (Reprinted with permission from Ref. 50. Copyright © 1984 Society for Applied Spectroscopy, Frederick, Maryland.)

GC/MS and GC/FT-IR chromatograms for this mixture are shown in Figures 6-25 and 6-26. GC/FT-IR was less sensitive than GC/MS and provided spectral information for only 75 of the 112 components. Most of the substances not detected by GC/FT-IR eluted toward the end of the separation. These substances were generally found to be high-molecular-weight PAHs that are poor infrared absorbers.

Search for alternate fuel sources has resulted in the development of coal-liquification processes. Coal liquification involves complex chemical reactions that are not well understood. Chemical analysis of coal-derived fluids is needed to better understand important processes involved in making liquid fuels from coal. Coal-derived fluids contain many nonvolatile and thermally labile organic species. For this reason, liquid chromatography is preferred for analysis of these mixtures. The enormous number of components in coal-derived fluids precludes complete

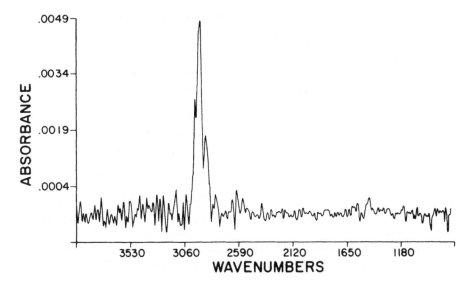

Figure 6-24 Reference GC/FT-IR spectrum of 40 ng of nonane injected on-column obtained by using the same apparatus and conditions as described for the jet fuel analysis. (Reprinted with permission from Ref. 50. Copyright © 1984 Society for Applied Spectroscopy, Frederick, Maryland.)

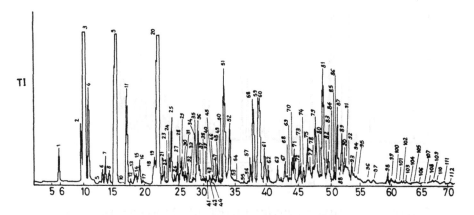

Figure 6-25 GC/MS total ion chromatogram for a petroleum distillate containing polycyclic aromatic hydrocarbons. (Reproduced from Ref. 54 by permission of Preston Publications, a division of Preston Industries, Inc., Niles, Illinois.)

Table 6-6   Library-Search Results for a Jet Fuel GC/FT-IR
Spectrum

Match number	Match factor	Library match
1	945	*m*-Xylene
2	1601	Bitolyl, *m*, mpr-,
3	1716	Hexane, 1,3,5-triphenyl-
4	1858	Benzene, 1-ethyl-3-methyl-,
5	1919	Benzene, 1,3,4-trimethyl-,
6	2004	Benzene, 1,3,5-trimethyl-,
7	2147	Benzene, 1,2,3,5-tetramethyl-,

Source: Reprinted with permission from Ref. 50. Copyright ©
1984 Society for Applied Spectroscopy, Frederick, Maryland.

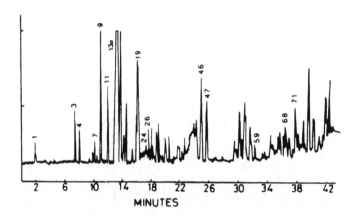

Figure 6-26   Gram–Schmidt GC/FT-IR chromatogram for a
petroleum distillate containing polycyclic aromatic hydrocarbons.
(Reproduced from Ref. 54 by permission of Preston Publications,
a division of Preston Industries, Inc., Niles, Illinois.)

separation by chromatographic techniques. Instead, these materials are fractionated into various classes prior to chromatographic analysis in order to reduce the number of components that must be separated. HPLC /FT-IR has been employed to obtain functional group information for various fractions from coal-derived fluids and process solvents [55—59].

Chloroform extracts from a solvent-refined coal fluid (SRC) were analyzed by using size-exclusion chromatography with FT-IR detection proposed by Brown et al. [55]. Chloroform extracts were fractionated into nine classes. Classes designated as basic nitrogen heterocycles (SESC 5), polyphenols (SESC 7), and components with high oxygen and nitrogen content (SESC 9) were then separated on a μ-Styragel HPLC column with chloroform mobile phase. Functional group-specific chromatograms for these separations are shown in Figure 6-27. The 2980 to 2850 cm^{-1} absorbance integration window was specific for aliphatic C—H detection, the 1630 to 1570 cm^{-1} window identified constituents containing aromatic rings, and the 1155 to 975 cm^{-1} window indicated the presence of ethers (C—O—C). Mixture components eluting early in the separations were those with the highest molecular weights. Molecular weight distributions for SESC 5 and SESC 7 were about the same, whereas the SESC 9 fraction contained primarily low-molecular-weight components. The SESC 5 fraction was found to have the highest concentration of ethers, many of which were high-molecular-weight species. The SESC 9 fraction also contained ethers, but these compounds were primarily low-molecular-weight substances.

Constituents of process solvents used for conversion of coal to coal fluids can dictate the composition of the liquefaction products. The effect of hydrotreating coal-derived fluids was studied by Brown and Taylor by using normal phase HPLC/ FT-IR [56]. The wide polarity range of components precluded complete separation of process solvents by normal phase chromatographic methods. Therefore, solvents were fractionated prior to analysis. An "intermediate polar" fraction was selected for separation on a Partisil-10 polar amino cyano (PAC) column by using CDCl$_3$ mobile phase. The use of the deuterated analog of chloroform permitted measurement of aliphatic C—H infrared absorptions. It was found that hydrotreating increases the ratio of aliphatic to aromatic C—H stretching absorbances, as would be expected for solvent hydrogenation. In addition, infrared spectral evidence indicated that ethers were converted to highly hindered phenols by hydrotreating.

VIII.   FLAVORS AND FRAGRANCES

Flavor and fragrance mixtures contain esters and acids for
which chromatography/FT-IR sensitivity is high.   Figure 6-28
is a GC/FT-IR-reconstructed chromatogram of a commercial
flavor additive.   More than 50 components were detected and
48 were characterized by library search.   Top search matches
for each of the 48 components are compiled in Table 6-7.   As
expected, library search results indicated the presence of
numerous esters.   However, inconsistences in search results
were observed.   For instance, top search identifications were
duplicated 5 times for the 48 detected components.   Chromatographic peaks 13 and 40 were both identified as cyclohexyl
butyrate.   Peaks 23 and 24 were identified as 1,1,1,3,3-penta-
chloro-2-propanone.   Peaks 33 and 46 were identified as

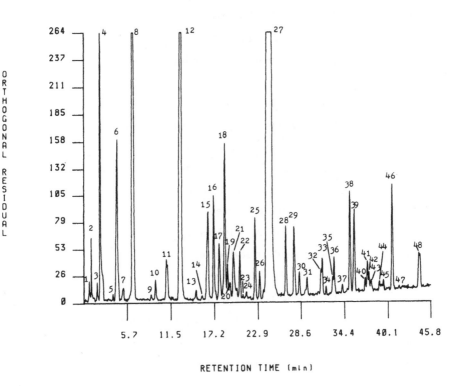

Figure 6-28   Gram–Schmidt GC/FT-IR chromatogram for a
flavor additive mixture.

Table 6-7   Top Search Matches for a Flavor Additive

Peak no.	Best search match
1	Distilled water
2	Ethyl alcohol
3	Acetic acid
4	Acetic acid, ethyl ester
5	Propionic acid, ethyl ester
6	1,2-Propanediol
7	Butyric acid, 2-bromo-3-methyl-,
8	Butyric acid, ethyl ester
9	Acetic acid, allyl ester
10	Acetic acid, butyl ester
11	Heptanoic acid, ethyl ester
12	Benzaldehyde
13	Butyric acid, cyclohexyl ester
14	2-Propanone, 1,1,3,3-tetrachloro-,
15	2-Buten-1-*ol*, 3-phenyl-,
16	Cyclohexene, 4-isopropenyl-1-methyl-,
17	Butyric acid, butyl ester
18	Benzaldehyde, *p*-methyl-,
19	Hexanoic acid, allyl ester
20	Lactic acid, butyl ester
21	Phenol, 2,4-dichloro-,
22	Isovaleric acid, isobutyl ester
23	2-Propanone, 1,1,1,3,3-pentachloro-,
24	2-Propanone, 1,1,1,3,3-pentachloro-,
25	Furfuryl alcohol, acetate

Table 6-7   (Continued)

Peak no.	Best search match
26	Succinic acid, acetyl-, diethyl ester
27	Anisole, P-bromo-,
28	P-Anisaldehyde
29	P-toluenesulfonic acid, butyl ester
30	3-Heptanol, 3,6-dimethyl-,
31	Hexanoic acid, cyclohexyl ester
32	Butyric acid, cinnamyl ester
33	Nonanoic acid, 4-hydroxy-, G-lactone
34	Glycidic acid, 3-phenyl-, ethyl ester
35	Cyclohexanol, 1-/3-/3-/diiso-propylamino/ethoxy/-1-PR
36	Benzaldehyde, 4-hydroxy-3-methoxy-,
37	Benzene, 1,2,4-tris/trimethyl-
38	Hexadecanedioic acid, dimethyl ester
39	Benzaldehyde, 4-hydroxy-3-methoxy-,
40	Butyric acid, cyclohexyl ester
41	1-Octen-3-ol
42	Benzene, 1,2-dimethoxy-4-propenyl-,
43	2-Propanone, hexachloro-,
44	Acrylic acid, allyl ester
45	2-butanone, 4-/P-hydroxyphenyl/-,
46	Nonanoic acid, 4-hydroxy-, G-lactone
47	2-Propanone, hexachloro-,
48	Citric acid, triethyl ester

4-hydroxy-G-lactone-nonanoic acid. Peaks 36 and 39 were identified as 4-hydroxy-3-methoxy-benzaldehyde, and peaks 43 and 47 as hexachloro-2-propanone. These duplications exemplify limitations of single-method detection and demonstrate a need for more comprehensive infrared libraries and additional structural information for unequivocal compound identification.

Integrated GC/FT-IR/MS has been employed to analyze a peppermint oil extract [60]. This sample was chosen to evaluate GC/FT-IR/MS performance during development of the instrumentation. As such, several of the mixture component identities were known prior to analysis. GC/MS total ion current and GC/FT-IR-integrated infrared absorbance chromatograms for this sample are shown in Figure 6-29, and search results for peppermint oil components are contained in Table 6-8. Of the 18 components analyzed, infrared spectral searches correctly identified 9 components and mass spectral searches identified 9 components. Seven of the 18 components were unambiguously identified by combined infrared and mass spectral search results.

Chromatography/FT-IR has been employed in a study to elucidate the nature of an animal response to a specific fragrance stimulus. Some Central and South American bees exhibit great affinity for specific tropical orchids [61]. These bees store fragrance material from the orchids in special organs in their rear legs. In order to study the relationship between the bees and the orchids, McClure et al. used GC/FT-IR to characterize orchid fragrances [62]. Extracted fragrance material was separated by using a 100 m × 0.32 mm bonded phase methylsilicone (5 μm film) capillary gas chromatograph column. Major components of the fragrance mixture were identified as myrcene, p-methylanisole, p-cymene, limonene, p-cresol, 1,8-cineole, 2-phenylethanol, ipsdienol, and a $C_{10}H_{14}$ olefin, for which there was no close library match. The unknown olefin was also found in a different orchid species and was believed to be a dehydration product of ipsdienol.

Naturally occurring fragrances often contain numerous components in widely varying concentrations. These materials may not be completely resolved by a single chromatographic separation. Fortunately, instrumental methods such as multidimensional gas chromatography can be used to overcome this problem. Slack and Heim applied multidimensional gas chromatography and FT-IR for analysis of a complex perfume sample

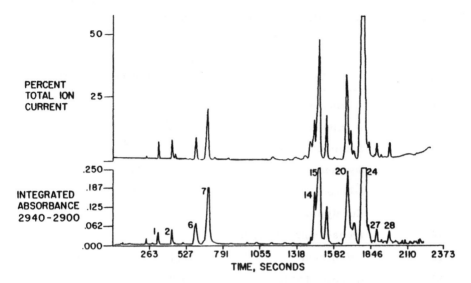

Figure 6-29   Results from GC/FT-IR/MS analysis of pepper-mint oil.   GC/MS chromatogram (top).   GC/FT-IR chromatogram (bottom).   (Reprinted with permission from Ref. 60. Copyright © 1982 American Chemical Society, Washington, D.C.)

[63].   Heart cutting was used to achieve separation of co-eluting components from a complex mixture.   A minor component of the heart cut was identified as alpha-terpineol.   The spectrum of this substance could not have been isolated without the additional chromatographic resolution provided by heart cutting and two-dimensional chromatography.

In another application of chromatography/FT-IR involving a naturally occurring fragrance, Herres et al. analyzed volatiles from cherimoya fruit (*Annona cherimolia, Mill.*) by GC/FT-IR [64].   Three fractions were obtained from 4 kg of pulp by high-vacuum distillation-liquid/liquid extraction, followed by liquid chromatographic separation.   Two of the three fractions were analyzed by capillary GC/FT-IR.   Results from these separations are compiled in Table 6-9.   Substance identifications listed in Table 6-9 were based on gas chromatographic retention-time comparisons and infrared spectral correlations with standards.   Vapor phase library searching was employed for additional identity confirmation.   Compounds listed in

Table 6-8   GC/FT-IR/MS Analysis of Peppermint Oil

Peak no.	Component	GC peak area	IR search	MS search
1	A-pinene	14.0	+	+
2	B-pinene	2.3	+	−
3	Sabinene	0.5	N[a]	−
4	Myrcene	0.3	N	+
6	l-Limonene	3.0	+	+
7	Eucalyptol	7.2	+	+
14	*trans*-Sabinene hydrate	1.0	−	−
15	l-Menthone	20.1	+	+
17	d-Isomenthone	5.7	+	+
20	Menthyl acetate	7.0	+	−
21	Neomenthol	3.6	+	+
22	Terpinene-4-*ol*	1.2	−	−
23	B-caryophyllene	2.8	−	−
24	l-Menthol	29.0	+	+
25	Pulegone	3.5	−	−
26	A-terpineol	0.8	−	−
27	Germacrene	2.0	−	−
28	Piperitone	1.3	−	+

[a]No infrared spectrum was obtained due to inadequate
sensitivity.

Table 6-9   Cherimoya Volatiles Identified by GC /FT-IR

Retention index

Sample	Reference	Compound
1007	1010	2-Methyl-1-propanol
1100	1121	1-Butanol
1105	1115	1-Penten-3-*ol*
1195	1177	3-Methyl-1-butanol
1215	1221	Butyl butanoate
1223	1217	(*E*)-3-penten-1-*ol*
1226	1221	1-Pentanol
1270	1267	3-Methylbutyl butanoate
1287	1287	3-Methylbutyl 3-methylbutanoate
1303	1305	Pentyl butanoate
1305	1294	(*Z*)-2-penten-1-*ol*
1324	1327	(*Z*)-2-pentenyl butanoate
1336	1334	Butyl 2-butenoate
1342	1331	1-Hexanol
1364	1359	(*Z*)-3-hexen-1-*ol*
1388	1383	(*E*)-2-hexen-1-*ol*
1415	1405	Hexyl butanoate
1428	1413	Acetic acid
1430	1434	Hexyl 3-methylbutanoate
1440	1444	3-Methylbutyl hexanoate
1500		Alcohol
1532	1532	Linalool

Table 6-9   (Continued)

Retention index		
Sample	Reference	Compound
1533		Unsaturated ketone
1576	1570	Dimethyl succinate
1617		Terpene alcohol
1674	1663	A-terpineol
1712	1700	1,4-Dimethoxybenzene
1716	1726	Myrtenol
1811	1803	Hexanoic acid
1814	1834	Benzyl butanoate
1836	1847	1-Undecanol

Source: Reprinted from Ref. 64. Copyright © 1983 Alfred Heuthig Publishers, New York.

Table 6-9 are believed to be the major volatile constituents of the fruit pulp. It was estimated that these substances are present at levels of $10-300$ µg/kg in fruit pulp.

IX. NATURAL PRODUCTS

In addition to GC/FT-IR analysis of naturally occurring fragrances, chromatography/FT-IR can be employed for analysis of other complex organic mixtures derived from living systems (natural products). Many components of these mixtures are isomers that are not readily identified by GC/MS. For example, Kakasinksy and McDonald detected and identified plant-derived terpenes by GC/FT-IR [65]. Terpene isomers are not readily identified by mass spectrometry. In fact, mass spectra of gamma-terpinene and alpha-pinene are virtually identical (Figure 6-30). In contrast, infrared spectra for these substances differ significantly (Figure 6-31). To evaluate GC/FT-IR for analysis of terpene-containing mixtures, Kalasinsky

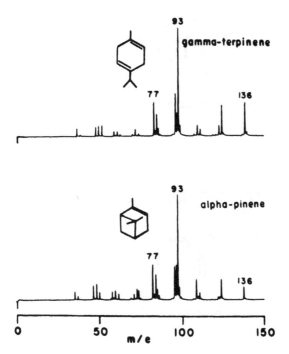

Figure 6-30 Mass spectra for gamma-terpinene and alpha-pinene. (Reproduced from Ref. 65 by permission of Preston Publications, a division of Preston Industries, Inc., Niles, Illinois.)

and McDonald studied the effects of chromatographic stationary phase on a separation of 12 terpenes dissolved in decane. Packed columns containing 10% Carbowax 20M and SE-30(2%)/OV-101(3%) were tested. Neither column provided complete separation of all 12 terpenes. However, by comparing infrared spectra obtained at the beginning and end of chromatographic elutions, overlapping elutions were identified. Interestingly, it was found that the technique of comparing terpene retention times with retention times of pure terpene standards dissolved in decane for identification of terpenes in the synthetic mixture was unreliable. This is illustrated by Table 6-10 that contains retention times (relative to decane) for separations of terpene mixture components and pure compounds for both of the columns employed. Some of the retention times were the

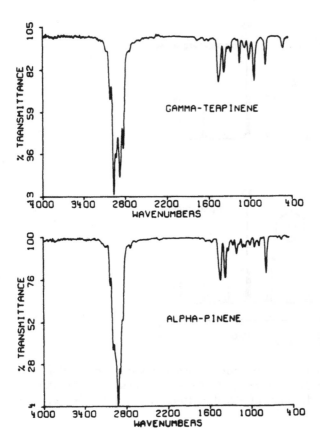

Figure 6-31   Vapor phase infrared spectra for gamma-terpinene and alpha-pinene. (Reproduced from Ref. 65 by permission of Preston Publications, a division of Preston Industries, Inc., Niles, Illinois.)

Table 6-10 Relative Retention Times (RRT) for Terpene Standards

Compound	SE-30/OV-101 column		Carbowax column	
	RRT[a] (binary)[b]	RRT (mixture)	RRT (binary)[b]	RRT (mixture)
a-Pinene	0.7	0.7	1.2	1.2
b-Pinene	0.8	0.9	1.7	1.7
a-Terpinene	1.0	1.1	2.2	2.2
p-Cymene	1.0	1.1	2.8	2.8
1,8-Cineole	1.1	1.1	2.4	2.4
Limonene	1.1	1.1	2.3	2.4
g-Terpinene	1.2	1.2	2.7	2.7
Fenchone	1.4	1.3	3.7	3.8
Camphor	1.6	1.6	4.6	4.7
Pulegone	2.2	2.1	5.7	5.6
d-Carvone	2.4	2.1	6.4	6.2
Thymol	2.4	2.4	8.8	8.7

[a]Relative to decane elution (decane RRT = 1.0).
[b]Binary solution of decane and pure terpene.
Source: Reproduced from Ref. 65 by permission of Preston Publications, a division of Preston Industries, Inc., Niles, Illinois.

same for pure component and mixture injections. However, several of the terpenes exhibited slightly different retention behavior when injected as a mixture compared to analysis of the pure substance. Therefore, it was important to obtain GC/FT-IR vapor phase spectra of these substances in order to provide additional information for use in making structural assignments.

Figure 6-32 contains integrated absorbance and Gram-Schmidt chromatograms for a pine tree extract known to contain terpene species. Although eight chromatographic elutions

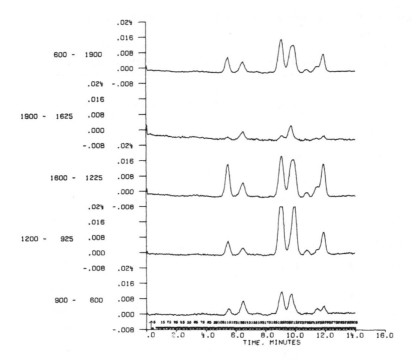

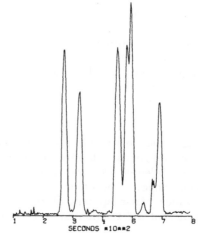

Figure 6-32 Chemigram (top) and Gram–Schmidt (bottom) GC/FT-IR chromatograms for a pine tree extract containing terpenes. (Reproduced from Ref. 65 by permission of Preston Publications, a division of Preston Industries, Inc., Niles, Illinois.)

can be discerned, nine components were identified. The first elution was found to contain two unresolved components. GC / FT-IR analysis with library-search spectral interpretation indicated that alpha-pinene, *t-p*-menthane, limonene, and 1,8-cineole were the major constituents of this mixture. Terpene analyses described thus far were performed using packed column gas chromatography. Improved chromatographic resolution was reported for capillary column chromatographic separations [66,67]. However, even with the added separating power of capillary chromatography, terpene mixtures may contain some components that are not completely separated [66−68]. To obtain more information for unresolved species by GC /FT-IR, separations with different columns can be employed and factor analysis used to predict the number of unresolved components based on infrared spectral comparisons.

The pine tree extract analyzed by Kalasinsky and McDonald was a relatively simple mixture. Terpene-containing samples often contain numerous components. For more complicated mixtures, multidimensional gas chromatography can be employed to simplify analysis. For example, Smith employed two-dimensional capillary gas chromatography to improve chromatographic separation of a complex terpene mixture [66]. Multidimensional detection can also be helpful in simplifying analysis of natural products for terpenes. For example, combined GC /MS and GC /FT-IR analysis of acid fractions isolated from sunflowers has been used to identify major and minor terpene components in these samples [68]. Figure 6-33 is an FID chromatogram of a terpene mixture separation obtained by using a DB-5 capillary column [69]. Areas of the chromatogram labeled as "heart cut" were diverted to a DB-WAX column and produced the chromatogram shown in Figure 6-34. The three unresolved chromatographic peaks selected as heart cuts were separated into more than 25 components by the DB-WAX column. Infrared spectra for some of the species represented in the chromatogram in Figure 6-34 are shown in Figure 6-35.

Purcell and Magidman analyzed coffee plant (*Coffea arabica*) aroma by GC /FT-IR in an attempt to detect the presence of microorganisms and insects [70]. Nitrogen was forced into a chamber containing coffee berries and passed through a TENAX-GC trap to concentrate aroma components. Samples were collected at 5 and 25°C. A specially designed desorption-injection apparatus was used to deposit aroma concentrates onto a Carbowax 20M packed column. Components were identified by infrared spectra and gas chromatographic retention-time

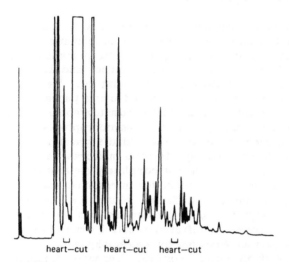

Figure 6-33   FID chromatogram for a terpene mixture.   Portions labeled "heart cut" were trapped and separated on a more polar capillary column.   (Reprinted from Ref. 69. Copyright © 1985 by International Scientific Communications, Inc., Fairfield, Connecticut.)

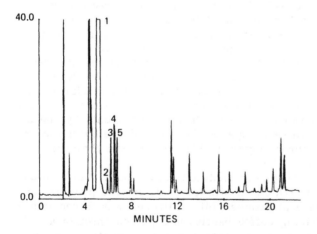

Figure 6-34   Gram–Schmidt GC/FT-IR chromatogram for the trapped species shown as heart cut in Figure 6-33.   Separation was achieved by using a Carbowax capillary column. (Reprinted from Ref. 69.   Copyright © 1985 by International Scientific Communications, Inc., Fairfield, Connecticut.)

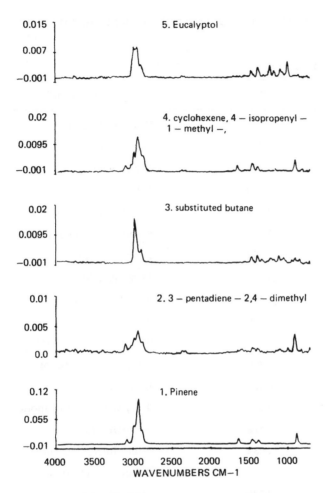

Figure 6-35 Vapor phase infrared absorbance spectra for chromatographic peaks numbered in Figure 6-34. (Reprinted from Ref. 69. Copyright © 1985 by International Scientific Communications, Inc., Fairfield, Connecticut.)

Table 6-11   Coffea Arabica Aroma Components Identified by
GC/FT-IR

Retention indices		
25°C	5°C	Compound
—	—	Carbon dioxide
946	947	Methyl acetate
995	991	Ethyl acetate
1016	1020	Ethanol
1074	1075	Ethyl propionate (tent.)
1133	1132	Ethyl butanoate (tent.)
1155	1154	Propyl acetate
1176	1177	Isovaleraldehyde
1200	1200	Hexyl acetate
1224	1220	A-phellandrene
—	—	Water
1273	1270	3-Methyl-1-butanol
1295	1294	Limonene
1392	1390	3-Hydroxy-2-butanone
1418	1418	G-terpinene
1432	1432	Phenylethyl acetate
1497	—	2-Octanol (tent.)
1530	1529	Linalool (tent.)

Source: Reprinted with permission from Ref. 70. Copyright ©
1984 Society for Applied Spectroscopy, Frederick, Maryland.

comparisons with standards. Table 6-11 contains a list of components identified in the aroma. Substances designated as tentatively identified were characterized by retention-time comparisons only. Carbon dioxide was ubiquitous in fruit aromas. Concentrations of CO_2 in excess of ambient arise from microorganism metabolism. A high CO_2 concentration therefore indicates the presence of agricultural contraband. Thus, infrared detection of carbon dioxide may provide a means of rapidly detecting agricultural contraband.

Analysis of phospholipids in tissues usually involves extraction of all lipids, followed by chromatographic separation and quantification by measuring phosphorus content. Chen and Kou developed an improved procedure for quantitative analysis of phospholipids in tissue based on HPLC/FT-IR [71]. Infrared analysis sensitivity was insufficient for monitoring minor phospholipid components, but was adequate for identifying and quantifying major components.

X. PHARMACEUTICALS

The use of structure specific GC/MS for drug analysis is well documented [72,73]. Recently, Kempfert evaluated GC/FT-IR as an analysis method for pharmaceutical products [74]. Figure 6-36 is a Gram – Schmidt GC/FT-IR chromatogram for a synthetic mixture of common "street drugs" exhibiting baseline separation of seven components. Vapor phase infrared spectra obtained from this separation were used to identify each of the drugs by library searching. Search results for peak 5 in Figure 6-36 are shown in Figure 6-37. The top search match was benzphetamine, which was in fact component 5 in the synthetic mixture. One advantage of GC/FT-IR compared to GC/MS is the capability of GC/FT-IR for differentiating drugs from their structurally similar metabolites. For example, vapor phase infrared spectra for lysergic acid diethyl amide (LSD) and lysergic acid methyl propyl amide (LAMPA) have significant differences in the fingerprint region (Figure 6-38). These species cannot be differentiated by mass spectrometric analysis. A wide variety of pharmaceutical products were studied by GC/FT-IR. Detection limits ranging from 50 to 100 ng for on-column injections were obtained for these substances. Gas chromatographic analysis of LSD presented a particularly difficult problem because this substance readily decomposes on hot metal surfaces. For this reason, on-column sample injection

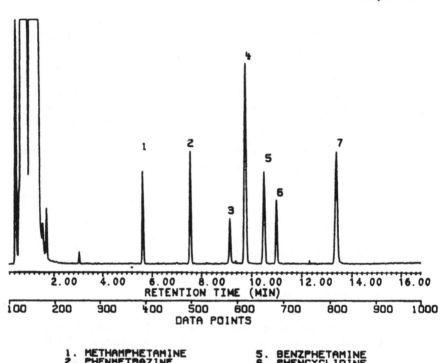

Figure 6-36   Gram–Schmidt GC/FT-IR chromatogram for a syn-
thetic mixture of "street drugs." (Reprinted from Ref. 74.
Copyright © 1988 Society for Applied Spectroscopy, Frederick,
Maryland.)

was employed and the capillary gas chromatographic column
was connected directly to the light pipe to prevent exposure
of eluting materials to metal surfaces.

GC/FT-IR has also been employed for analysis of drugs
in equine urine [75]. Hordenine was isolated from equine
urine by extraction, followed by thin-layer chromatography.
A scrape was made to remove hordenine from the TLC plate.
The drug was then dissolved in ethyl acetate and injected

PEAK #5

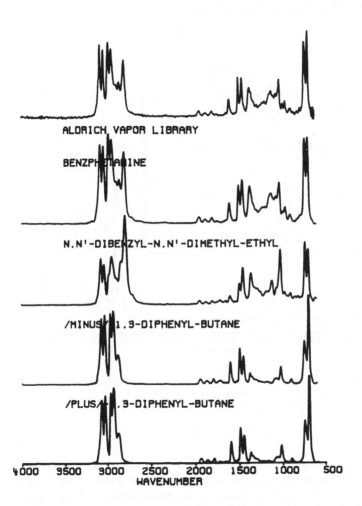

Figure 6-37 Vapor phase infrared library-search results for chromatographic elution 5 shown in Figure 6-36. (Reprinted from Ref. 74. Copyright © 1988 Society for Applied Spectroscopy, Frederick, Maryland.)

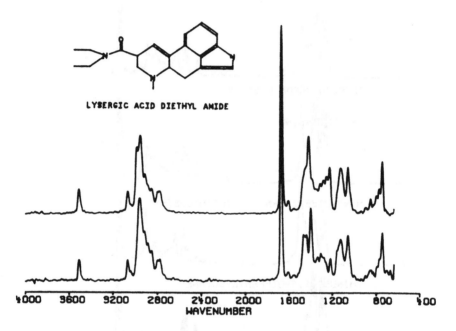

Figure 6-38   Vapor phase infrared spectra of LSD (top) and
LAMPA (bottom). (Reprinted from Ref. 74. Copyright © 1988
Society for Applied Spectroscopy, Frederick, Maryland.)

into a GC/FT-IR for analysis.   TLC separation alone was sufficient for separating hordenine from other urine components.
Therefore, the analysis could have been performed by TLC/
FT-IR without the additional sample preparation for GC/FT-IR.
In fact, TLC/FT-IR analysis of pharmaceuticals has been reported [76,77]. Detection limits in the high nanogram range
have been attained for separations requiring about 20 min.
    Aqueous solution analysis of choline pharmaceuticals was
performed by using an attenuated total reflectance (CIRCLE

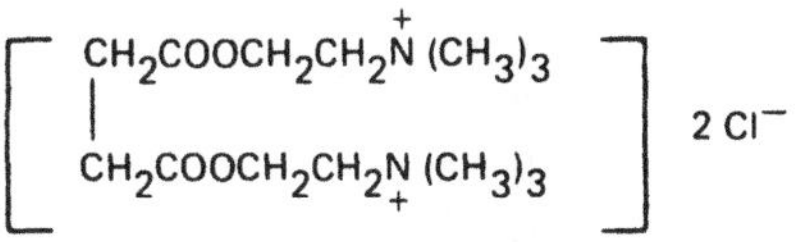

$$\left[\begin{array}{l} CH_2COOCH_2CH_2\overset{+}{N}\,(CH_3)_3 \\ | \\ CH_2COOCH_2CH_2\underset{+}{N}\,(CH_3)_3 \end{array} \right] \; 2\,Cl^-$$

(a)

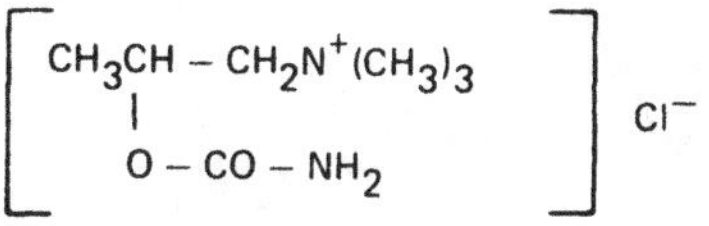

$$\left[\begin{array}{l} CH_3CH - CH_2N^+(CH_3)_3 \\ | \\ O - CO - NH_2 \end{array} \right] \; Cl^-$$

(b)

Figure 6-39   Molecular structures for (a) succinylcholine chloride and (b) bethanechol chloride. (Reprinted from Ref. 78. Copyright © 1988 Society for Applied Spectroscopy, Frederick, Maryland.)

cell) HPLC /FT-IR interface [78]. Flow injection analysis (FIA) with FT-IR detection was employed as a quick analysis scheme for succinylcholine chloride and bethanechol chloride. HPLC is preferred to GC in this application because these pharmaceuticals do not have sufficient volatility and therefore must be derivatized prior to GC analysis [79,80]. Structures for these two pharmaceuticals are shown in Figure 6-39. Succinylcholine chloride is a fast-acting muscle relaxant and bethanechol chloride a cholerigenic drug. Aqueous phase infrared spectra for these substances are shown in Figure 6-40. A Dowex 2-X8 strong anion exchange resin HPLC column was used for flow injection separations. Quantitative analysis of the two drugs was obtained by measuring the infrared absorbance for these species at 953 cm^{-1} (trimethyl quaternary ammonium band) or 1165 cm^{-1} (C —O —C stretching vibration) for succinylcholine chloride and 1075 cm^{-1} (C —O —C stretching vibration) for bethanechol chloride. Detection limits as low as 40 μg, on-column were obtained, yielding linear dynamic ranges from 1 to 100 ppt for each substance.

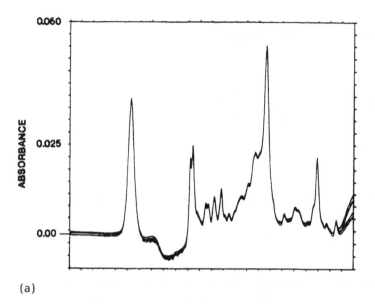

(a)

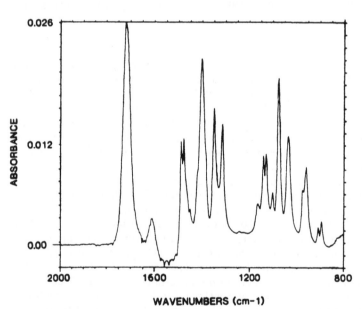

(b)

Figure 6-40   Aqueous phase infrared spectra for (a) a 40 ppt solution of succinylcholine chloride (5 replicates superimposed) and (b) 20 ppt of bethanechol chloride obtained by FIA/FT-IR. (Reprinted from Ref. 78.   Copyright © 1988 Society for Applied Spectroscopy, Frederick, Maryland.)

XI. INDUSTRIAL APPLICATIONS

Witt et al. described one of the first industrial applications of
GC /FT-IR for analysis of impurities in a commercial solvent
[81]. Commercial-grade divinyl benzene (DVB) is commonly
used as a polymerizing monomer. The reactivity of DVB is
dependent on the impurities it contains. GC /FT-IR was used
to identify the major components of DVB as *m*-ethylvinyl,
p-ethylvinyl, and divinyl benzene. GC /FT-IR analysis indicated that low-level quantities of *p*-diethyl benzene, *o*-diethyl
benzene, 1,2,3,4-tetrahydronaphthalene, *o*-DVB, ethylallyl
benzene, and vinylallyl benzene were also present in commercial DVB. Structure assignments for these components were
confirmed by mass spectrometric measurements.

Several other applications of chromatography /FT-IR to
polymer industry problems have been reported [48,82-84].
For example, Luoma and Rowland used GC /FT-IR and HPLC /
FT-IR to study the aging process of acrylic structural adhesives [82]. These substances contain a variety of different
polymeric materials that may react during aging to change the
physical properties of the polymer. GC /FT-IR was used to
identify the volatile components of polymers tested. Typical
constituents of these mixtures included methyl-methacrylate,
2-methylacrylic acid, 2-hydroxyethyl methacrylic acid,
N,N-dimethyl aniline, ethylene glycol dimethacrylate, and
2-methyl-*m*-phenyl isocyanic acid.

Capillary SFC with carbon dioxide mobile phase can
separate nonvolatile, thermally labile polymer additives with
high resolution more readily than HPLC. Raynor et al.
described an SFC /FT-IR buffer-memory interface in which
polymer additives were deposited on an infrared transparent
window and analyzed by infrared microscopy [85]. Twenty-
one different polymer additives were studied by SFC /FT-IR.
Trade names and chemical names for these substances are
listed in Table 6-12. Chemical structure and elution order for
these species are given in Table 6-13. All but one of the
polymer additives studied were solid at room temperature and
therefore could be deposited at ambient temperature without
sample loss. A polypropylene extract analysis was demonstrated using infrared spectral correlations and retention-time
comparisons to identify various additives in the polymer.

Structural information provided by GC /FT-IR was employed to characterize the hydrogenation reaction of trifluoroethylene [86]. The desired product of hydrogenation was

Table 6-12   Polymer Additives Studied by SFC/FT-IR

Peak no.	Trade name	Chemical name
1	Topanol OC	2,4,6-tri-*tert*-butylphenol
2	Tinuvin P	2-(2-hydroxy-5-methylphenyl)-2*H*-benzotriazole
3	Tinuvin 292	bis(1-methyl-2,2,6,6-tetra-methylpiperidinyl) sebacate
4	Tinuvin 320	2-(2-hydroxy-3,5-di-*tert*-butyl-phenyl)-2*H*-benzotriazole
5	Tinuvin 326	2-(3-*tert*-butyl-2-hydroxy-5-methylphenyl)-2*H*-5-chloro-benzotriazole
6	Tinuvin 328	2-(2-hydroxy-3,5-di-*tert*-amyl-phenyl)-2*H*-benzotriazole
7	Chimassorb 81	2-hydroxy-4-*n*-octyloxybenzo-phenone
8	Erucamide	*cis*-13-docosenamide
9	Tinuvin 770	bis(2,2,6,6-tetramethyl-4-piperi-dinyl) sebacate
10	Tinuvin 440	8-acetyl-3-dodecyl-7,7,9,9-tetra-methyl-1,3,8-triazaspiro(4,5)-decane-2,4-dione
11	Irgafos 168	tris(2,4-di-*tert*-butylphenyl) phosphite
12	Tinuvin 144	2-*tert*-butyl-2-(4-hydroxy-3,5-di-*tert*-butylbenzyl)[bis-(methyl-2,2,6,6-tetra-methyl-4-piperidinyl)] dipropionate
13	Irganox PS800	dilauryl thiodipropionate
14	Irganox 1076	octadecyl-3-(3,5-di-*tert*-butyl-4-hydroxyphenyl) propionate
15	Irganox MD1025	*N*,*N*-bis[1-oxo-3-(3,5-di-*tert*-butyl-4-hydroxyphenyl) propane]hydrazine

Table 6-12 (Continued)

Peak no.	Trade name	Chemical name
16	Irganox 245	triethylene glycol bis-3-(3-*tert*-butyl-4-hydroxy-5-methyl-phenyl)propionate
17	Irganox 1035	2,2-thiodiethylene bis[3-(3,5-di-*tert*-butyl-4-hydroxy-phenyl)propionate
18	Irganox 3114	tris(3,5-di-*tert*-butyl-4-hydroxybenzyl) isocyanurate
19	Irganox PS802	distearyl thiodipropionate
20	Irganox 1330	1,3,5-tris(3,5-di-*tert*-butyl-4-hydroxybenzyl)-2,4,6-trimethylbenzene
21	Irganox 1010	pentaerythritol tetrakis [3-(3,5-di-*tert*-butyl-4-hydroxyphenyl)propionate]

Source: Reprinted with permission from Ref. 85. Copyright © 1988 American Chemical Society, Washington, D.C.

1,1,2-trifluoroethane. GC /FT-IR chromatograms for the reaction mixture revealed three components. The first component was an unsaturated fluorocarbon and found to be unreacted trifluoroethylene. The second chromatographic component was identified as 1,1-difluoroethane that was a reaction byproduct. The third component was identified as the desired 1,1,2-trifluoroethane product.

As stated previously, GC /FT-IR is a powerful tool for the analysis of vapors. It is therefore useful in industry for rapid qualitative analysis of vapors and for research and development studies of volatile commercial products. For example, Killar et al. investigated head-space GC /FT-IR for use in petroleum-industry applications [87]. Soils were analyzed for hydrocarbon contamination and alcohols were detected in gasoline by using this technique. Identifiable spectra were obtained for 0.02% (by volume) methanol concentrations in gasoline.

Table 6-13   Chemical Structures and Elution Order of Polymer Additives

Peak No	Chemical Structure	Peak No	Chemical Structure	Peak No	Chemical Structure
1		10		18	
2		11		19	
3		12		20	
4		13		21	
5		14			
6		15			
7		16			
8		17			
9					

Source: Reprinted with permission from Ref. 85.   Copyright © 1988 American Chemical Society, Washington, D.C.

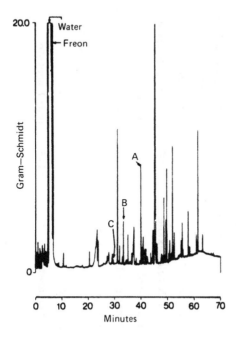

Figure 6-41 Gram – Schmidt GC/FT-IR chromatogram for a 100 μl injection of Freon-11 containing chocolate volatiles. (Reprinted with permission from Ref. 88. Copyright © 1986 American Chemical Society, Washington, D.C.)

An injector/trap capillary gas chromatograph sample introduction device was used by Fehl and Marcott to identify chocolate volatiles [88]. Chocolate aroma was isolated by steam distillation-extraction with Freon-11. The Gram – Schmidt-reconstructed chromatogram for a 100 μl injection of extracted volatiles is shown in Figure 6-41. More than 85 interpretable infrared spectra were obtained. Chromatographic peaks labeled A, B, and C in Figure 6-41 were identified as phenethyl acetate, tetramethylpyrazine, and trimethylpyrazine, respectively. All three are known constituents of chocolate. Wilkins et al. reported analysis of a commercial lacquer thinner by GC/FT-IR/MS [60]. Thirty components were detected by both infrared and mass spectrometry. Table 6-14 contains search comparisons for the 30 components. Molecular weight

Table 6-14 GC/FT-IR/MS Analysis of a Lacquer Thinner Sample

Peak	Compound type[a]	FT-MS[b]	FT-IR[b]	CI mol. wt.	Identity
1	$CH_3C(=O)R$	1-dimethylaminoacetone	acetone	58	acetone
2			methanol	32	methanol
3	Sec. alcohol	2-propanol	2-propanol	60	2-propanol
4	$CH_3C(=O)R$	3,4-dimethyl-2-hexanone	2-butanone	72	2-butanone
5	Tert. alcohol	3,7-dimethyl-3-octanol	3,7-dimethyl-3-octanol		3,7-dimethyl-3-octanol
6	Sub. alkane	3-methylhexane	3,3-dimethylpentane		
7	Sub. cyclopentane	1,1-dimethylcyclopentane	methylcyclopentane	98	1,1-dimethylcyclopentane
8	Alkane	heptane	hexane		heptane
9	Sub. cyclohexane	methylcyclohexane	methylcyclohexane	98	methylcyclohexane
10		sec-butyl acetate	2-butanol	74	2-butanol
11	Alcohol	3-methyl-1-butanol	3-methyl-1-pentanol	102	3-methyl-1-pentanol
12	Sub. cyclopentane	1,1,2-trimethyl-cyclopentane	butylcyclopentane	112	1,1,2-trimethyl-cyclopentane
13	Sub. benzene	toluene	toluene	92	toluene
14		2,2-dimethyl-1,3-propanediol	butyl acetate	116	butyl acetate
15	Sub. alkane	2,3,3,4-tetramethyl-pentane	2,4-dimethylhexane		2,4-dimethylhexane
16		6-decen-5-one	2,5-dimethyl-2-hexene	112	2,5-dimethyl-2-hexene

No.	Compound type[a]				
17	Large alkane	2-(5-cyclohexyl)-undecane	2,6,10,14-tetramethyl-pentadecane		
18	Sub. cycloalkane	1,1,3,4-tetramethyl-cyclopentane	cis-1,3-dimethylcyclo-hexane	126	1,1,3,4-tetramethyl-cyclopentane
19	Alkene	2,3,6-trimethyl-4-octene	2,5-dimethyl-2-hexene		
20	Sub. alkane	3-methylheptane	3,3-dimethylheptane	114	3-methylheptane
21	Sub. alkane	3,4-dimethylheptane	3,3-dimethylheptane		dimethylheptane
22	Sub. alkane	3-ethylheptane	3-ethylpentane		
23			2,4-dimethylhexane		
24	Sub. alkane	2,2,4-trimethylheptane	2,2,4-trimethylhexane		
25		xylene	butylcyclopentane	106	xylene
26	Alkane	3,3-dimethylhexane	octane		
27		xylene	2-butoxyethyl acetate	106	xylene
28	Bicyclic alkane	bicyclo[3.1.0]hex-2-ene, 2-methyl-5-(1-methyl-ethyl)-	bicyclo[3.1.1]hept-2-ene, 2,6,6-trimethyl-	136	
29	2-ethoxy ester	2-ethoxyethyl acetate	2-ethoxyethyl acetate	132	2-ethoxyethyl acetate
30		bicyclo[3.1.0]hexane, 4-methylene-1-(1-methylethyl)			

[a]Compound type is the structural feature most common in the two search output lists.

[b]The specific compounds listed are not those with the highest similarity index on their respective searches. Instead, they are the compounds considered to be most similar to one another when the search lists are compared.

Source: Reprinted with permission from Ref. 60. Copyright © 1982 American Chemical Society, Washington, D.C.

information for 18 of the 30 components was derived from chemical ionization mass spectrometry. In the four cases in which infrared and mass spectrometric search results agreed, unequivocal identification was supported by molecular weight estimates derived from chemical ionization mass spectrometry. Of the 30 components jointly detected by infrared and mass spectrometry, structural information was obtained for 28 and 18 were identified unequivocally.

XII. CONCLUSION

It is clear from the chromatography/FT-IR applications described here that in most instances infrared spectroscopy alone cannot provide unequivocal mixture-component identification. For this reason, chromatography/FT-IR results are often combined with retention indices or mass spectral analysis to improve structure assignments. FT-IR can provide isomer differentiation, but usually cannot distinguish homologues. On the other hand, MS can distinguish homologues, but sometimes cannot distinguish isomers. The GC/FT-IR/MS combination has proven to be a useful complex mixture analysis technique that *can* provide unequivocal and unambiguous compound identification. Differences in FT-IR and MS sensitivity have limited joint analysis applications to components present in moderate to high concentrations. Clearly, there is a need for increased chromatography/FT-IR sensitivity to extend infrared analysis to trace mixture components. Techniques such as matrix isolation and subambient trapping provide GC/FT-IR detection limits in the mid-picogram range. Light pipe interfaces are unable to match trapping interfaces in sensitivity. As a result, future GC/FT-IR/MS combinations will probably incorporate modifications of trapping interfaces available today. With commercial availability of low-cost infrared and mass spectrometers, GC/FT-IR/MS will likely become the method of choice for volatile complex mixture analysis in the near future. HPLC/FT-IR, SFC/FT-IR, and TLC/FT-IR are not as sensitive as GC/FT-IR, but are more appropriate for analyses involving nonvolatile mixture components. Potential benefits derived from these combinations more than justify future efforts to improve interface performances.

REFERENCES

1. W. Worthy, *Chem. and Eng. News, 65* (36): 33 (1987).

2. D. F. Gurka, L. D. Betowski, T. L. Jones, S. M. Pyle, R. Titus, J. M. Ballard, Y. Tondeur, and W. Niederhut, *J. Chromatogr. Sci., 26*: 301 (1988).

3. W. L. Budde and J. W. Eichelberger, *Anal. Chem., 51*: 567A (1979).

4. D. Rosenthal, *Anal. Chem., 54*: 63 (1982).

5. D. F. Gurka and R. Titus, *Anal. Chem., 58*: 2189 (1986).

6. D. F. Gurka, M. H. Hiatt, and R. L. Titus, *Hazardous and Industrial Solid Waste Testing: Fourth Symposium, ASTM STP*, Vol. 886 (J. K. Petros, Jr., W. J. Lacy, and R. A. Conway, eds.), American Society for Testing and Materials, Philadelphia, Pa., p. 139 (1986).

7. D. F. Gurka, *Appl. Spectrosc., 39*: 827 (1985).

8. D. F. Gurka, P. R. Laska, and R. Titus, *J. Chromatogr. Sci., 20*: 145 (1982).

9. D. F. Gurka, R. Titus, P. R. Griffiths, D. Henry, and A. Giorgetti, *Anal. Chem., 59*: 2362 (1987).

10. D. F. Gurka and S. M. Pyle, *Environ. Sci. Technol., 22*: 963 (1988).

11. K. H. Shafer, M. Cooke, F. DeRoos, R. J. Jakobsen, O. Rosario, and J. D. Mulik, *Appl. Spectrosc., 35*: 469 (1981).

12. D. F. Gurka, M. Umana, E. D. Pellizzari, A. Moseley, and J. A. de Haseth, *Appl. Spectrosc., 39*: 297 (1985).

13. G. R. Umbreit, *Chromatographic Analysis of the Environment* (R. L. Grob, ed.), Marcel Dekker, New York, p. 85 (1983).

14. M. M. Mossoba, J. T. Chen, W. C. Brumley, and S. W. Page, *Anal. Chem., 60*: 945 (1988).

15. A. Nettleship, P. S. Henshaw, and H. L. Meyer, *J. Natl. Cancer Inst., 4*: 309 (1943).

16. W. M. Shaub and W. Tsang, *Environ. Sci. Technol.*, *17*: 721 (1983).

17. T. J. Nestrick, L. L. Lamparski, and R. H. Stehl, *Anal. Chem.*, *51*: 2273 (1979).

18. R. K. Mitchum, G. F. Moler, and W. A. Korfmacher, *Anal. Chem.*, *52*: 2278 (1980).

19. H. R. Buser, *Anal. Chem.*, *48*: 1553 (1986).

20. C. J. Wurrey, S. Bourne, and R. D. Kleopfer, *Anal. Chem.*, *58*: 482 (1986).

21. D. F. Gurka, J. W. Brasch, R. H. Barnes, C. J. Riggle, and S. Bourne, *Appl. Spectrosc.*, *40*: 978 (1986).

22. J. Grainger, E. Barnhart, D. G. Patterson, Jr., and D. Presser, *Appl. Spectrosc.*, *42*: 321 (1988).

23. T. T. Holloway, B. J. Fairless, C. E. Freidline, H. E. Kimball, R. D. Kloepfer, C. J. Wurrey, L. A. Jonooby, and H. G. Palmer, *Appl. Spectrosc.*, *42*: 359 (1988).

24. J. Grainger, V. V. Reddy, and D. G. Patterson, Jr., *Appl. Spectrosc.*, *42*: 800 (1988).

25. K. H. Shafer, S. V. Lucas, and R. J. Jakobsen, *J. Chromatogr. Sci.*, *17*. 464 (1979).

26. K. H. Shafer, A. Bjorseth, J. Tabor, and R. J. Jakobsen, *J. High Res. Chromatogr. Chromatogr. Commun.*, *3*: 87 (1980).

27. L. V. Azarraga and C. A. Potter, *J. High Res. Chromatogr. Chromatogr. Commun.*, *4*: 60 (1981).

28. H. Malissa, Jr., G. Szolgyenyi, and K. Winsauer, *Fresenius Z. Anal. Chem.*, *321* 17 (1985).

29. D. F. Gurka and L. D. Betowski, *Anal. Chem.*, *54*: 1819 (1982).

30. D. F. Gurka, M. Hiatt, and R. Titus, *Anal. Chem.*, *56*: 1102 (1984).

31. D. F. Gurka and R. Titus, *1985 International Conference on Fourier and Computerized Infrared Spectroscopy*, Vol. 553 (J. G. Grasselli and D. G. Cameron, eds.), The International Society for Optical Engineering, Bellingham, Wash., p. 236 (1985).

32. K. H. Shafer, T. L. Hayes, J. W. Brasch, and R. J. Jakobsen, *Anal. Chem.*, *56*: 237 (1984).

33. M. D. Erickson, D. L. Newton, E. P. Pellizzari, K. B. Tomer, and D. Dropkin, *J. Chromatogr. Sci.*, *17*: 449 (1979).

34. D. R. Latham, C. R. Ferrin, and J. S. Ball, *Anal. Chem.*, *34*: 311 (1962).

35. J. H. Bell, S. Ireland, and A. W. Spears, *Anal. Chem.*, *41*: 310 (1969).

36. D. Schuetzle, F. S. Lee, T. J. Prater, and S. B. Tejada, *Int. J. Environ, Anal. Chem.*, *2*: 93 (1981).

37. J. N. Pitts, *Philos. Trans. Roy. Soc. London, Ser. A.*, *290*: 551 (1979).

38. H. S. Rosenkranz, *Mutat. Res.*, *140*: 1 (1984).

39. I. Saleem, A. M. Durison, T. J. Prater, T. Riley, and D. Schuetzle, *Mutat. Res.*, *104*: 17 (1982).

40. V. F. Kalasinsky, C. Saiwan, and K. G. Whitehead, *J. Chromatogr. Sci.*, *26*: 584 (1988).

41. M. D. Erickson, S. D. Cooper, C. M. Sparacino, and R. A. Zweidinger, *Appl. Spectrosc.*, *33*: 575 (1979).

42. M. D. Erickson, *Appl. Spectrosc.*, *35*: 181 (1981).

43. K. S. Kalasinsky, *1981 International Conference on Fourier and Computerized Infrared Spectroscopy*, Vol. 289 (H. Sakai, ed.), The International Society for Optical Engineering, Bellingham, Wash., p. 156 (1981).

44. K. S. Kalasinsky, *J. Chromatogr. Sci.*, *21*: 246 (1983).

45. K. S. Kalasinsky, *Chemical, Biological and Industrial Applications of Infrared Spectroscopy* (J. R. Durig, ed.), Wiley, Chichester, England, p. 277 (1985).

46. W. N. Sanders and J. B. Maynard, *Anal. Chem.*, *40*: 527 (1968).

47. S. E. Garlock, G. E. Adams, and S. L. Smith, *Am. Lab. (Fairfield, Conn.)*, *14* (12): 48 (1982).

48. S. L. Smith, S. E. Garlock, and G. E. Adams, *Appl. Spectrosc.*, *37*: 192 (1983).

49. J. Hubball, R. Raborvsky, S. Gerosa, and J. Criscio, *Am. Lab.*, *12* (10): 121 (1980).

50. J. R. Cooper and L. T. Taylor, *Appl. Spectrosc.*, *38*: 366 (1984).

51. P. J. Duff, *1985 International Conference on Fourier and Computerized Infrared Spectroscopy*, Vol. 553 (J. G. Grasselli and D. G. Cameron, eds.), The International Society for Optical Engineering, Bellingham, Wash., p. 118 (1985).

52. P. E. Eberly, Jr., C. N. Kimberlin, Jr., W. H. Miller, and A. V. Drushel, *Ind. Eng. Chem. Process Des. Dev.*, *5*: 193 (1966).

53. I. Mochida, K. Otsuka, K. Maeda, and K. Takeshita, *Carbon*, *13*: 135 (1975).

54. V. N. Garg, B. D. Bhatt, V. K. Kaushik, and K. R. Murthy, *J. Chromatogr. Sci.*, *25*: 237 (1987).

55. R. S. Brown, D. W. Hausler, and L. T. Taylor, *Anal. Chem.*, *53*: 197 (1981).

56. R. S. Brown and L. T. Taylor, *Anal. Chem.*, *55*: 723 (1983).

57. R. S. Brown and L. T. Taylor, *Anal. Chem.*, *55*: 1492 (1983).

58. P. G. Amateis and L. T. Taylor, *Chromatographia, 18*: 175 (1984).

59. P. G. Amateis and L. T. Taylor, *Anal. Chem.*, *56*: 966 (1984).

60. C. L. Wilkins, G. N. Giss, R. L. White, G. M. Brissey, and E. C. Onyiriuka, *Anal. Chem.*, *54*: 2260 (1982).

61. N. H. Williams and W. M. Whitten, *Biol. Bull.*, *164*: 355 (1983).

62. G. L. McClure, N.H. Williams, and W. M. Whitten, *1985 International Conference on Fourier and Computerized Infrared Spectroscopy*, Vol. 553 (J. G. Grasselli and D. G. Cameron, eds.), The International Society for Optical Engineering, Bellingham, Wash., p. 355 (1985).

63. R. W. Slack and A. C. Heim, *Am. Lab. (Fairfield, Conn.)*, *18* (8): 80 (1986).

64. W. Herres, H. Idstein, and P. Schreier, *J. High Res. Chromatogr. Chromatogr. Commun.*, *6*: 590 (1983).

65. V. F. Kalasinsky and J. T. McDonald, Jr., *J. Chromatogr. Sci.*, *21*: 193 (1983).

66. S. L. Smith, *Appl. Spectrosc.*, *40*: 278 (1986).

67. V. F. Kalasinsky, S. Pechsiri, and K. S. Kalasinsky, *J. Chromatogr. Sci.*, *24*: 543 (1986).

68. S. L. Smith, *J. Chromatogr. Sci.*, *22*: 143 (1984).

69. S. L. Smith, *Am. Lab. (Fairfield, Conn.)*, *17* (11): 82 (1985).

70. J. M. Purcell and P. Magidman, *Appl. Spectrosc.*, *38*: 181 (1984).

71. S. S. Chen and A. Y. Kou, *J. Chromatogr.*, *307*: 261 (1984).

72. W. A. Garland, *J. Pharm. Sci.*, *66*: 77 (1977).

73. L. K. Pickering, J. L. Hoecker, and W. G. Kramer, *Clin. Chem.*, *25*: 300 (1979).

74. K. Kempfert, *Appl. Spectrosc.*, *42*: 845 (1988).

75. G. A. Maylin, E. A. Dewey, and J. D. Henion, *LC-GC*, 5 (10): 904 (1987).

76. H. Bui, *Spectroscopy*, 2 (10): 44 (1987).

77. K. H. Shafer, J. A. Herman, and H. Bui, *Am. Lab. (Fairfield, Conn.)*, *20* (2): 142 (1988).

78. B. E. Miller, N. D. Danielson, and J. E. Katon, *Appl. Spectrosc.*, *42*: 401 (1988).

79. J. Balkon, B. Donnelly, and T. A. Rejent, *J. Anal. Toxicol.*, *7*: 237 (1983).

80. R. B. Forney, Jr., F. T. Carroll, I. K. Nordgren, B. B. Pettersson, and B. Holmstedt, *J. Anal. Toxicol.*, *6*: 115 (1982).

81. J. D. Witt, M. K. Gabriel, and R. L. Julian, *J. Chromatogr. Sci.*, *17*: 445 (1979).

82. G. A. Luoma and R. D. Rowland, *J. Chromatogr. Sci.*, *24*: 210 (1986).

83. S. L. Smith, *J. Chromatogr.*, *279*: 623 (1983).

84. W. Herres and T. Dee, *1985 International Conference on Fourier and Computerized Infrared Spectroscopy*, Vol. 553 (J. G. Grasselli and D. G. Cameron, eds.), The International Society for Optical Engineering, Bellingham, Wash., p. 337 (1985).

85. M. W. Raynor, K. D. Bartle, I. L. Davies, A. Williams, A. A. Clifford, J. M. Chalmers, and B. W. Cook, *Anal. Chem.*, *60*: 427 (1988).

86. J. M. Chalmers, M. W. Mackenzi, and H. A. Willis, *Appl. Spectrosc.*, *38*: 763 (1984).

87. R. G. Kollar, R. Citerin, M. Markelov, and K. L. Gallaher, *1985 International Conference on Fourier and Computerized Infrared Spectroscopy*, Vol. 553 (J. G. Grasselli and D. G. Cameron, eds.), The International Society for Optical Engineering, Bellingham, Wash., p. 192 (1985).

88. A. J. Fehl and C. Marcott, *Anal. Chem.*, *58*: 2578 (1986).

Index

A

Absorptivity, 69, 100, 140, 163
Acenapthene, 32
Acetoaminophen, 152
Acetone, 122, 188
Acetophenone, 32, 107, 108
Acetylsalicylic acid, 145, 151, 155
Adhesives, 307
Adsorption chromatography, 1, 95
Airborne species, 263
Alanine, 147
Alpha-pinene, 292, 293, 294
Alumina 139-141
American Petroleum Institute (API), 164
American Society for Testing and Materials (ASTM), 164

Ames test, 264
Amino acids, 147
Analog-to-digital converter, 14
Aniline, 257
Anisole, 108, 111
Anti-5, 10-dihydrogen mirex, 269, 272
Arginine, 147
Array processor, 23, 187
Artificial intelligence, 218
Asbestos, 236
Attenuated total reflectance, 44, 112, 304

B

Basis set, 24, 25
Basis vector, 24, 25
Beam profiling, 61
Beer's law, 35, 98, 107, 140
Benzphetamine, 301, 303